New BASIC SKILLS WITH MATH

General Math Review

New BASIC SKILLS WITH MATH
General Math Review

JERRY HOWETT

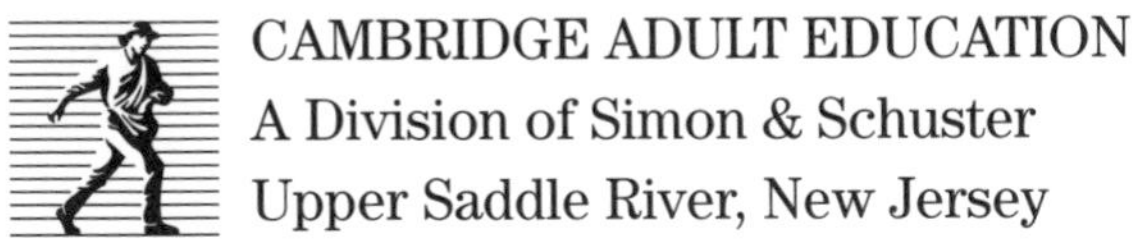

CAMBRIDGE ADULT EDUCATION
A Division of Simon & Schuster
Upper Saddle River, New Jersey

Editorial Developer: Catherine Fillmore Hoyt

Executive Editor: Mark Moscowitz

Editors: Karen Bernhaut, Doug Falk, Amy Jolin, Kristen Shepos-Salvatore

Marketing Manager: Will Jarred

Production Director: Penny Gibson

Production Editor: Alan Dalgleish

Book Design: Parallelogram, New York

Electronic Page Production: Burmar Technical Corporation, Albertson, New York

Cover Art: Salem Krieger

Cover Design: Pat Smythe

Printed in the United States of America

1 2 3 4 5 6 7 8 9 10 99 98 97 96 95

ISBN 0-8359-4658-4

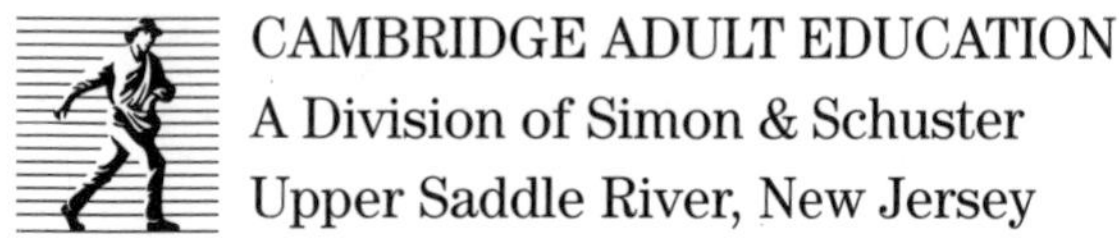

Contents

BASIC SKILLS PREVIEW

These problems will help you find out which parts of this book you need to work on. Do all the problems you can. At the end of the problems, look at the chart to see which page you should go to next.

Solve each problem.

1. Write *two hundred forty thousand, three hundred twelve* in figures.

2. What is 29,703 rounded to the nearest thousand?

3. $5{,}382 + 678 + 13{,}209 =$

4. $18{,}004 - 9{,}256 =$

5. $628 \times 56 =$

6. $1{,}083 \times 27 =$

7. Sean can drive 28 miles on one gallon of gasoline. How far can he drive if his car holds 12 gallons of gasoline?

8. $962 \div 26 =$

9. $2{,}167 \div 9 =$

10. Mark is saving $125 a month to buy a boat. How many months will it take him to save $2,750?

11. Write *eighty and fifty-two hundredths* in figures.

12. Which decimal is greatest, .54, .504, or .45?

13. Round .1287 to the nearest thousandth.

14. $.076 + 3.24 + 21 =$

15. $12 - .607 =$

16. $.0038 \times 62 =$

17. Phil bought 2.8 pounds of cheese. The cheese cost $3.80 a pound. How much did Phil pay for the cheese?

18. $46.4 \div 16 =$

19. $148 \div 3.7 =$

20. Harry paid $13.05 for 4.5 feet of lumber. Find the price per foot.

21. Reduce $\frac{6}{54}$ to lowest terms.

22. Change $\frac{24}{10}$ to a mixed number and reduce.

23. Change $5\frac{2}{3}$ to an improper fraction.

24. Which is greater, $\frac{7}{12}$ or $\frac{5}{8}$?

25. $3\frac{1}{2} + 1\frac{3}{4} + 2\frac{7}{10} =$

26. $7\frac{4}{9} - 4\frac{5}{6} =$

27. $15 \times 1\frac{2}{5} =$

28. $2\frac{1}{4} \times 4\frac{2}{3} =$

29. Lois bought $3\frac{1}{4}$ pounds of chicken for $1.20 a pound. How much did she pay for the chicken?

30. $\frac{5}{12} \div 10 =$

31. $3\frac{1}{2} \div 2\frac{4}{5} =$

32. Jack cut a board $60\frac{3}{4}$ inches long into three equal pieces. How long was each piece?

33. Simplify the ratio 24:32.

34. The Conrads spend $480 on rent and $1,440 on everything else each month. What is the ratio of their rent to the total amount they spend each month?

35. Write and simplify the ratio of 800 pounds to one ton.

36. Solve for n in the proportion $n:3 = 200:9$.

37. Shirley drove 126 miles in three hours. Driving at the same rate, how far can she go in five hours?

38. The ratio of newcomers to returning students at the Midtown Adult Learning Center is 3:10. Altogether there are 380 students at the center. How many of them are newcomers?

39. Change .6 to a percent.

40. Change $\frac{3}{10}$ to a percent.

41. Change 36% to a fraction.

42. 225% of 48 =

43. $12\frac{1}{2}$% of 136 =

44. Kate bought a vest on sale. The vest originally cost $34, but it was on sale for 15% off. How much did Kate pay for the vest?

45. Find the interest on \$920 at 6% annual interest for one year and three months.

46. Last year the Gomez family paid \$450 a month for rent. This year they have to pay \$477 a month. By what percent did their rent go up?

47. 25% of what number is 35?

48. 20% of the students in Mr. Green's math class were absent because of a snowstorm on Thursday. Five students were absent. How many students are in the class?

49. $8^2 - 5^2 =$

50. $\sqrt{16} + \sqrt{9} =$

51. $-12 - 9 =$

52. $(-9)(-10) =$

53. $\frac{-15}{-20} =$

54. $3(12 - 7) =$

55. Find the value of $7x - 2y$ when $x = 4$ and $y = 3$.

56. Write an algebraic expression for fifteen less than n.

57. Solve for w in $8w - 1 = 95$.

58. Nine less than four times a number is eleven. Find the number.

59. Solve for a in $7a = 48 + 3a$.

60. Solve for n in $3(n - 1) = 12$.

61. In the illustration, $\angle x = 58°$. Find the measurement of $\angle y$.

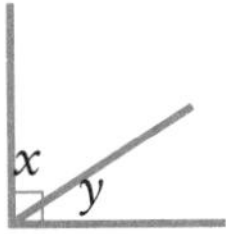

62. In the illustration, $\angle m = 113°$. Find the measurement of $\angle n$.

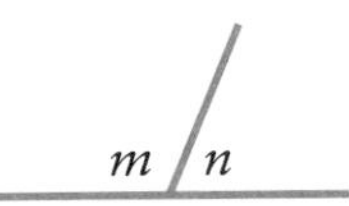

For problems 63–64, find the perimeter of each figure.

63. $w = 16$ in.
$l = 24$ in.

64. 16 ft.
13 ft. 13 ft.

For problems 65–66, find the area of each figure.

65. $s = 14$ ft

66. $h = 12$ yd
$b = 27$ yd

67. Find the volume of a rectangular container that is 10 feet long, 5 feet wide and 4 feet high.

68. The two triangles in the illustration are similar. Find the measurement of side x.

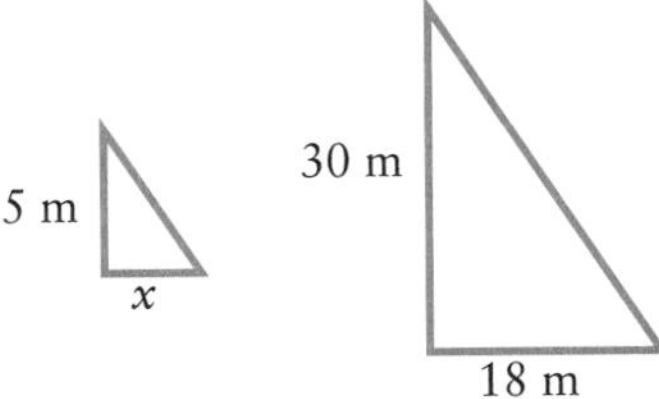

PROGRESS CHECK

Check your answers on page 173. Then complete the chart below.

Problem numbers	*Number of problems in this section*	*Number of problems you got right in this section*	
1 to 10	10	______________	If you had fewer than 8 problems right, go to page 7.
11 to 20	10	______________	If you had fewer than 8 problems right, go to page 30.
21 to 32	12	______________	If you had fewer than 10 problems right, go to page 51.
33 to 38	6	______________	If you had fewer than 4 problems right, go to page 78.
39 to 48	10	______________	If you had fewer than 8 problems right, go to page 90.
49 to 60	12	______________	If you had fewer than 10 problems right, go to page 114.
61 to 68	8	______________	If you had fewer than 6 problems right, go to page 143.

UNIT 1

Whole Numbers

Place Value

Whole numbers are made up of the **digits** 0, 1, 2, 3, 4, 5, 6, 7, 8, and 9. The number 44 has two digits. The number 23,060 has five digits. The value of each digit is different because of its position in the number. Every position has a **place value**. The table below gives the names of the first ten places in our whole number system.

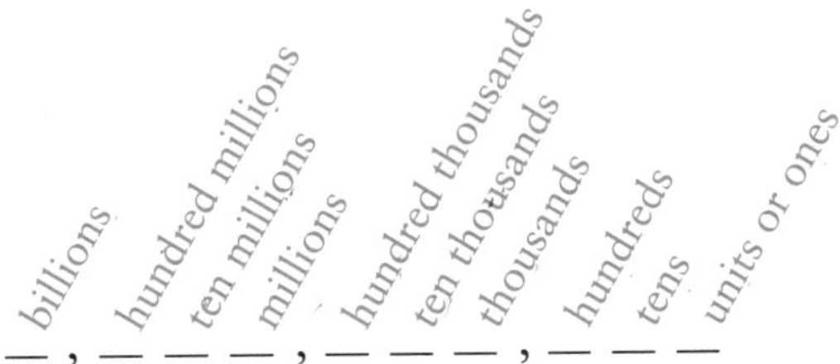

Example: Find the value of 5 in 85,406.

The digit 5 is in the thousands place. It has a value of 5 thousands or 5,000.

PRACTICE 1

Use the number 985 to answer problems 1–3. Item 1 is done for you.

1. 9 is in the _hundreds_ place. 9 has a value of _900_.

2. 8 is in the ______ place. 8 has a value of ______.

3. 5 is in the ______ place. 5 has a value of ______ .

There are 1,760 yards a mile. Use this number to answer problems 4–6.

4. 1 is in the ______ place. 1 has a value of ______.

5. 7 is in the ______ place. 7 has a value of ______.

6. 6 is in the ______ place. 6 has a value of ______.

To check your answers, turn to page 173.

Reading and Writing Whole Numbers

Commas make numbers easier to read. Counting from the right, there is a comma after every three places. Large numbers are read in groups of three. At each comma we say the name of the group of digits that are set off by the comma.

PRACTICE 2

A. Supply the missing words you need to read each number.

Example: 2,043,000 two _million_, forty-three _thousand_.

Notice how the commas go in the same place whether the number is written in words or figures.

1. 502 five ______ two.

2. 4,080 four ________, eighty.

3. 58,320 fifty-eight ________, three hundred twenty.

4. 6,019,300 six ________, nineteen ________, three hundred.

5. 246,500 two hundred forty-six ________, five hundred.

Write words to show how to read each number.

6. 3,800 ______________________________

7. 19,007,200 ______________________________

To write whole numbers from words, watch for places that must be held with zeros.

Example: Write three million, four hundred eight thousand, six hundred as a whole number.

This number contains no ten thousands, no tens, and no units. Hold these places with zeros.

3,408,600

B. Write each of the following as a whole number.

8. three hundred eight ____________

9. two hundred sixty-one thousand ____________

10. ninety thousand, twenty-four ____________

11. four million, one hundred seventy thousand ________________

12. eight hundred four thousand, five hundred ________________

13. sixty thousand, three hundred ________________

14. eleven million, two hundred seven thousand ________________.

To check your answers, turn to page 174.

Rounding Whole Numbers

Cory's car weighs 2,374 pounds. What is the weight of her truck to the nearest 100 pounds?

To find the weight to the nearest 100 pounds, you must *round off* 2,374. Rounding makes numbers easier to use when you don't need exact values.

To round a whole number:

1. Underline the digit in the place to which you want to round.
2. If the digit to the right of the underlined digit is more than 4, add 1 to the underlined digit.
3. If the digit to the right of the underlined digit is less than 5, do not change the underlined digit.
4. Replace the digits to the right of the underlined digit with zeros.

Example: Round 2,374 to the nearest hundred.

STEP 1.	Underline the digit in the hundreds place, 3.	$2,\underline{3}74$
STEP 2.	The digit to the right of 3 is 7. Add 1 to 3, and replace 6 and 2 with zeros.	2,400

➡ To the nearest 100 pounds, Cory's car weighs 2,400 pounds.

Example: Round 29,624 to the nearest thousand.

Step 1. Underline the digit in the thousands place, 9. 2<u>9</u>,624

Step 2. The digit to the right of 9 is 6. Add 1 to 9. Since 1 + 9 = 10, you must carry 1 over to the ten-thousands column. Replace 6, 2, and 4 with 0s. 30,000

PRACTICE 3

Round each number.

1. to the nearest ten:	38	542	295	5,291	108
2. to the nearest hundred:	863	59,848	4,082	647	3,954
3. to the nearest thousand:	6,174	39,723	3,279	20,736	

To check your answers, turn to page 174.

Addition with Carrying

At the ABM Assembly plant there are 857 workers on the day-shift and 268 on the night shift. Find the total number of workers at ABM.

The answer to an addition problem is called the **sum** or **total.** When the sum of the digits in a column is a two-digit number, *carry* the digit at the left to the next column to the left.

Example: 857 + 268 =

Step 1. Add the ones. 7 + 8 = 15. Write 5 under the ones column, and carry the 1 to the tens column. The 1 is one ten. It must be added to the tens column.

```
   1
  857
+ 268
-----
    5
```

STEP 2. Add the tens. $1 + 5 = 6$ and $6 + 6 = 12$. Write the 2 under the tens and carry the 1 to the hundreds.

STEP 3. Add the hundreds. $1 + 8 = 9$ and $9 + 2 = 11$.

$$\begin{array}{r} {\scriptstyle 1\,1} \\ 857 \\ +\ 268 \\ \hline 1{,}125 \end{array}$$

➡ The total number of workers at ABM is 1,125.

PRACTICE 4

Add each problem

1. $\begin{array}{r} 44 \\ +\ 57 \\ \hline \end{array}$ $\begin{array}{r} 78 \\ +\ 24 \\ \hline \end{array}$ $\begin{array}{r} 37 \\ +\ 63 \\ \hline \end{array}$ $\begin{array}{r} 91 \\ +\ 89 \\ \hline \end{array}$ $\begin{array}{r} 46 \\ +\ 98 \\ \hline \end{array}$ $\begin{array}{r} 73 \\ +\ 28 \\ \hline \end{array}$ $\begin{array}{r} 68 \\ +\ 35 \\ \hline \end{array}$

2. $\begin{array}{r} 15 \\ +\ 88 \\ \hline \end{array}$ $\begin{array}{r} 74 \\ +\ 56 \\ \hline \end{array}$ $\begin{array}{r} 63 \\ +\ 79 \\ \hline \end{array}$ $\begin{array}{r} 84 \\ +\ 28 \\ \hline \end{array}$ $\begin{array}{r} 97 \\ +\ 36 \\ \hline \end{array}$ $\begin{array}{r} 46 \\ +\ 99 \\ \hline \end{array}$ $\begin{array}{r} 67 \\ +\ 63 \\ \hline \end{array}$

3. $\begin{array}{r} 68 \\ +\ 46 \\ \hline \end{array}$ $\begin{array}{r} 87 \\ +\ 75 \\ \hline \end{array}$ $\begin{array}{r} 89 \\ +\ 52 \\ \hline \end{array}$ $\begin{array}{r} 53 \\ +\ 27 \\ \hline \end{array}$ $\begin{array}{r} 62 \\ +\ 48 \\ \hline \end{array}$ $\begin{array}{r} 18 \\ +\ 55 \\ \hline \end{array}$ $\begin{array}{r} 77 \\ +\ 66 \\ \hline \end{array}$

4. $\begin{array}{r} 341 \\ +\ 59 \\ \hline \end{array}$ $\begin{array}{r} 228 \\ +\ 85 \\ \hline \end{array}$ $\begin{array}{r} 368 \\ +\ 92 \\ \hline \end{array}$ $\begin{array}{r} 625 \\ +\ 77 \\ \hline \end{array}$ $\begin{array}{r} 439 \\ +\ 46 \\ \hline \end{array}$ $\begin{array}{r} 773 \\ +\ 96 \\ \hline \end{array}$

5. $\begin{array}{r} 48 \\ +\ 485 \\ \hline \end{array}$ $\begin{array}{r} 92 \\ +\ 378 \\ \hline \end{array}$ $\begin{array}{r} 57 \\ +\ 974 \\ \hline \end{array}$ $\begin{array}{r} 65 \\ +\ 573 \\ \hline \end{array}$ $\begin{array}{r} 83 \\ +\ 267 \\ \hline \end{array}$ $\begin{array}{r} 24 \\ +\ 388 \\ \hline \end{array}$

6. $\begin{array}{r} 775 \\ +\ 638 \\ \hline \end{array}$ $\begin{array}{r} 593 \\ +\ 549 \\ \hline \end{array}$ $\begin{array}{r} 206 \\ +\ 297 \\ \hline \end{array}$ $\begin{array}{r} 335 \\ +\ 866 \\ \hline \end{array}$ $\begin{array}{r} 184 \\ +\ 499 \\ \hline \end{array}$ $\begin{array}{r} 288 \\ +\ 827 \\ \hline \end{array}$

Rewrite and add each problem.

7. 739 + 494 = 867 + 584 = 766 + 804 =

8. 932 + 1,478 = 2,551 + 488 = 368 + 2,944 =

9. 6,544 + 2,476 = 3,982 + 1,077 = 1,256 + 4,855 =

10. 10,649 + 23,288 = 53,279 + 17,072 =

11. The distance from New York to Cleveland is 507 miles. The distance from Cleveland to Chicago is 343 miles. What is the distance from New York to Chicago by way of Cleveland?

12. In a recent election for mayor, Mr. Green got 4,987 votes and Mr. Munro got 4,062 votes. Find the total number of votes cast for these two candidates.

13. Sam's truck weighs 3,478 pounds. If he loads it with 1,800 pounds of topsoil, what is the the combined weight of the truck and the topsoil?

14. The Johnsons pay $548 a month for rent and $176 a month on their car loan. What is the total of these monthly expenses?

15. In June volunteer firefighters raised $16,479 toward the purchase of a new vehicle. In July they raised another $14,208. Altogether how much money did they raise in June and July?

To check your answers, turn to page 174.

Addition of More Than Two Numbers

Carlos works in the shipping department of an electrical supply company. For one customer he packed three boxes. One weighed 29 pounds, the second weighed 14 pounds, and the third weighed 7 pounds. Find the total weight of the three boxes.

To add more than two numbers, find the total for each column. The digits in a column can be added in any order.

Example: 29 + 14 + 7 =

STEP 1. Line up the numbers and add the digits in the units column. 9 + 4 = 13. Then 13 + 7 = 20. Write 0 in the units column and carry 2 to the tens.

$$\begin{array}{r} {}^{2} \\ 29 \\ 14 \\ +\ 7 \\ \hline 50 \end{array} \quad (9+4=13,\ 13+7=20)$$

STEP 2. Add the digits in the tens column. 2 + 2 = 4. Then 4 + 1 = 5.

➠ The total weight of the boxes is 50 pounds.

PRACTICE 5

Solve.

1.

17	44	38	47	39	24	43
88	60	27	96	53	25	44
+ 53	+ 67	+ 69	+ 80	+ 58	+ 73	+ 19

2.

26	45	59	88	66	56	12
55	32	80	47	25	63	84
+ 66	+ 29	+ 41	+ 70	+ 63	+ 34	+ 57

3.

236	808	375	686	327
1,940	2,767	3,086	6,421	8,448
+ 375	+ 741	+ 829	+ 506	+ 338

4.

58	23	77	94	79
964	158	841	349	926
4,277	3,284	1,624	6,953	1,756
+ 52	+ 41	+ 12	+ 54	+ 88

Rewrite and solve.

5. 318 + 9,907 + 24,063 = 7,613 + 24 + 88,552 =

6. 8,016 + 11,238 + 127 = 43 + 1,752 + 18,406 =

7. 79,088 + 314 + 2,607 = 935 + 22,463 + 8,142 =

8. In March Don's Music Shop sold 1,026 tapes, in April they sold 963 tapes, and in May they sold 1,372 tapes. What were the total sales for those three months?

9. For lunch Manny had a bowl of chicken soup (207 calories), a ham sandwich (324 calories), coffee with cream (30 calories), and a piece of apple pie (330 calories). What was the total number of calories in his lunch?

10. Gordon paid $115 for new brakes, $19 for a new shock absorber, and $18 for an oil change. Find the total of these items.

To check your answers, turn to page 175.

Subtraction with Regrouping

Carpenters at the ABM Company have to build 85 packing crates. By Wednesday they had finished 39 crates. How many do they have left?

The answer to a subtraction problem is called the **difference.** When the bottom number in any column is too large to subtract from the top number, you must **regroup** the top number. You may know this operation as *renaming* or *borrowing.*

Example: 85 − 39 =

You can also show regrouping like this:

$$\begin{array}{r} {}^{7}\!\not{8}{}^{1}5 \\ -\ 39 \\ \hline 46 \end{array}$$

STEP 1. 9 is too large to subtract from 5. Regroup the top number. Take 1 ten from the tens column (8 − 1 = 7) and add it to the units (10 + 5 = 15).

$$\begin{array}{r} 7\ 15 \\ \not{8}\not{5} \\ -\ 39 \\ \hline 46 \end{array}$$

STEP 2. Subtract the units. 15 − 9 = 6.

STEP 3. Subtract the tens. 6 − 5 = 1.

The carpenters have 46 crates left to build.

PRACTICE 6

Subtract each problem.

1. 58 − 29 96 − 48 47 − 28 58 − 19 51 − 43 32 − 16 93 − 65

$$\begin{array}{r} 58 \\ -\ 29 \\ \hline \end{array} \quad \begin{array}{r} 96 \\ -\ 48 \\ \hline \end{array} \quad \begin{array}{r} 47 \\ -\ 28 \\ \hline \end{array} \quad \begin{array}{r} 58 \\ -\ 19 \\ \hline \end{array} \quad \begin{array}{r} 51 \\ -\ 43 \\ \hline \end{array} \quad \begin{array}{r} 32 \\ -\ 16 \\ \hline \end{array} \quad \begin{array}{r} 93 \\ -\ 65 \\ \hline \end{array}$$

2.

$$\begin{array}{r} 52 \\ -\ 25 \\ \hline \end{array} \quad \begin{array}{r} 64 \\ -\ 38 \\ \hline \end{array} \quad \begin{array}{r} 25 \\ -\ 16 \\ \hline \end{array} \quad \begin{array}{r} 87 \\ -\ 18 \\ \hline \end{array} \quad \begin{array}{r} 91 \\ -\ 55 \\ \hline \end{array} \quad \begin{array}{r} 33 \\ -\ 16 \\ \hline \end{array} \quad \begin{array}{r} 46 \\ -\ 28 \\ \hline \end{array}$$

3.

$$\begin{array}{r} 811 \\ -\ 243 \\ \hline \end{array} \quad \begin{array}{r} 572 \\ -\ 418 \\ \hline \end{array} \quad \begin{array}{r} 467 \\ -\ 199 \\ \hline \end{array} \quad \begin{array}{r} 340 \\ -\ 238 \\ \hline \end{array} \quad \begin{array}{r} 551 \\ -\ 365 \\ \hline \end{array} \quad \begin{array}{r} 760 \\ -\ 467 \\ \hline \end{array}$$

4.

6,175	7,240	5,628	6,425	3,183
− 496	− 384	− 979	− 556	− 287

5.

4,236	5,668	6,673	4,290	3,837
− 1,448	− 2,699	− 3,887	− 2,947	− 2,608

Rewrite each problem with ones under ones, tens under tens, and so on. Then subtract.

6. 564 − 85 = 414 − 56 = 666 − 79 =

7. 5,335 − 2,914 = 2,414 − 1,671 = 7,380 − 4,093 =

8. 15,624 − 9,587 = 34,124 − 5,025 = 86,472 − 9,913 =

9. 683 people signed up to go on a trip to Miami. 506 people actually went on the trip. How many people who signed up did not go?

10. The Garcias must drive 413 miles to get to their son's house. They stopped to eat lunch after they had driven 225 miles. How much farther did they have to drive?

11. Sam Brown earns $23,246 a year. Jane Brown earns $29,175 a year. How much more does Jane make in a year than Sam?

12. The Globe Theatre holds 420 people. At a Saturday night show, 87 seats were empty. How many people were at the show that night?

To check your answers, turn to page 175.

Regrouping with Zeros

Imani had $904 in her savings account. She withdrew $356 to buy new winter clothes. How much did she have left in her account?

To regroup with zeros, look at the first digit in the top number that is not zero.

Example: 904 − 356 =

STEP 1. You cannot subtract 6 from 4. Take 1 hundred from the hundreds column (9 − 1 = 8). You now have 10 tens in the tens column.

```
 8 10
 9 0 4
− 3 5 6
```

STEP 2. Take 1 ten from the 10 in the tens column (10 − 1 = 9) and add it to the units (10 + 4 = 14).

```
    9
 8 10 14
 9  0  4
− 3  5  6
  5  4  8
```

STEP 3. Subtract the units. 14 − 6 = 8.

STEP 4. Subtract the tens. 9 − 5 = 4.

STEP 5. Subtract the hundreds. 8 − 3 = 5.

➡ Imani had $548 left in her account.

PRACTICE 7

Subtract each problem.

1. 801 − 236 407 − 209 503 − 388 808 − 619 506 − 218 707 − 379

2. 706 − 267 503 − 125 802 − 708 901 − 355 704 − 477 206 − 168

3.

600	800	500	900	300	400
− 226	− 354	− 189	− 614	− 108	− 376

4.

4,000	3,000	8,000	2,000	7,000	6,000
− 1,256	− 2,338	− 4,411	− 1,950	− 3,076	− 2,447

Rewrite and subtract each problem.

5. 7,000 − 1,270 =   3,000 − 1,681 =   9,000 − 4,023 =

6. 18,007 − 5,668 =   20,050 − 9,266 =   30,600 − 9,482 =

7. 90,040 − 18,255 =   60,000 − 13,478 =   40,005 − 20,386 =

8. The town of Midvale wants to raise $850,000 to build a new health center. They have collected $473,260 so far. How much more money do they need?

9. The Martinez family bought a house for $128,000. They made a down payment of $ 19,200. How much more do they owe for the house?

10. Frank borrowed $2,800 to buy a used car. So far he has paid back $1,675. How much more does Frank owe on the loan?

11. The U.S. produces 7,120,000 barrels of crude oil each day. Iraq produces 450,000 barrels a day. In one day, the U. S. produces how much more oil than Iraq?

To check your answers, turn to page 175.

Multiplication with Carrying

The shipping department at Ace Electrical sent out 29 crates of motor parts. Each crate weighed 76 pounds. Find the total weight of the shipment.

The answer to a multiplication problem is called the *product*. When you multiply two digits, the product is often a two-digit number. You must **carry** the left digit to the next number you are multiplying. Then **add** the digit you carry to the next product.

Example: 76 × 29 =

STEP 1. 9 × 6 = 54. Write 4 in the units column, and carry 5 to the next column.

STEP 2. 9 × 7 = 63. Add the 5 that you carried. 63 + 5 = 68.

STEP 3. 2 × 6 = 12. Write 2 in the tens column, carry the 1 to the next column.

STEP 4. 2 × 7 = 14. Add the 1 that you carried. 14 + 1 = 15.

STEP 5. Add the partial products.

```
   76
 × 29
 ----
  684 ← partial products
 152  ←
 ----
 2204 ← product
```

➡ The total weight of the shipment was 2,204 pounds.

PRACTICE 8

Multiply each problem.

1. 74 × 6 82 × 9 37 × 8 68 × 5 39 × 4 22 × 9 54 × 3

2. 36 × 38 42 × 65 25 × 43 84 × 36 77 × 46 64 × 55 33 × 78

Rewrite each problem. Put the number with fewer digits on the bottom. Then multiply.

3. 778 × 63 = 46 × 684 = 277 × 28 = 82 × 756 =

4. 24 × 866 = 679 × 34 = 73 × 492 = 536 × 57 =

5. 48 × 30 = 75 × 50 = 60 × 167 = 20 × 396 =

6. Rodney makes $370 a week. There are 52 weeks in a year. How much does Rodney make in one year?

7. One gallon of paint costs $16. Find the price of seven gallons of paint.

8. Donell can drive 27 miles on one gallon of gasoline. How far can he drive on 12 gallons of gasoline?

9. There are 12 inches in one foot. How many inches long is a board that measures 15 feet?

10. Marcella drove at an average speed of 64 mph for three hours. How far did she drive? [**Hint:** Multiply her driving rate by the time she drove.]

11. Jose makes $14 an hour. How much does he earn in a week if he works 35 hours?

12. Mark is paying back a car loan. He has to pay $160 a month for 24 months. Find the total amount he is paying back.

To check your answers, turn to page 176.

Division by One Digit

Serena has a part-time job at a convenience store. She makes $6 an hour. Last month she made $504. How many hours did she work?

The answer to a division problem is called the **quotient.** To find a quotient, repeat the four steps listed below until you complete the problem.

1. Divide.
2. Multiply.
3. Subtract and compare.
4. Bring down the next number.

Example: Divide 6 into 504.

STEP 1. **Divide:** 6 goes into 50 eight times. Write 8 above the tens place.

$$\begin{array}{r} 8 \\ 6\overline{)504} \end{array}$$

STEP 2. **Multiply:** $8 \times 6 = 48$. Write 48 under 50.

STEP 3. **Subtract:** $50 - 48 = 2$. **Compare** to be sure that what you get by subtraction is less than what you divide by. 2 is less than 6.

$$\begin{array}{r} 8 \\ 6\overline{)504} \\ \underline{48} \\ 2 \end{array}$$

STEP 4. **Bring down the next number:** 4.

$$\begin{array}{r} 8 \\ 6\overline{)504} \\ \underline{48} \\ 24 \end{array}$$

STEP 5. **Divide:** 6 goes into 24 four times. Write 4 above the units place.

$$\begin{array}{r} 84 \\ 6\overline{)504} \\ \underline{48} \\ 24 \end{array}$$

STEP 6. **Multiply:** $4 \times 6 = 24$. Write 24 under 24.

STEP 7. **Subtract:** $24 - 24 = 0$. **Compare:** 0 is less than 6.

$$\begin{array}{r} 84 \\ 6\overline{)504} \\ \underline{48} \\ 24 \\ \underline{24} \\ 0 \end{array}$$

➡ Serena worked 56 hours.

Writing every step, as the example shows in step 7, is called **long division.** In **short division** you write only the answer and the number you get by subtracting.

Example: $\begin{array}{r} 8\ 4 \\ 6\overline{)50^{2}4} \end{array}$

PRACTICE 9

Divide each problem.

1. $3\overline{)141}$ $9\overline{)207}$ $2\overline{)170}$ $5\overline{)280}$ $7\overline{)308}$

2. $8\overline{)616}$ $7\overline{)252}$ $6\overline{)270}$ $5\overline{)345}$ $4\overline{)348}$

3. $3\overline{)237}$ $6\overline{)348}$ $5\overline{)320}$ $8\overline{)704}$ $3\overline{)276}$

4. $9\overline{)2,484}$ $2\overline{)1,678}$ $7\overline{)2,562}$ $3\overline{)2,553}$ $6\overline{)5,568}$

5. Three friends equally shared a raffle prize of $750. How much did each of them get?

6. The Simpsons paid $13,800 for a five-acre piece of land. What was the price of one acre?

To check your answers, turn to page 176.

Division with Remainders

How many buckets, each containing 4 quarts of paint, can Alan fill from a vat that contains 387 quarts of paint?

If you do not get zero in the last subtraction step of a division problem, you will have a **remainder.**

Example:

$$\begin{array}{r} 96\text{r}3 \\ 4\overline{)387} \\ \underline{36} \\ 27 \\ \underline{24} \\ 3 \end{array}$$

➡ Alan can fill 96 buckets. There will be 3 quarts left over.

PRACTICE 10

A. Divide each problem.

1. $7\overline{)292}$ $2\overline{)79}$ $9\overline{)204}$ $6\overline{)316}$ $3\overline{)236}$

2. $4\overline{)243}$ $3\overline{)169}$ $8\overline{)357}$ $7\overline{)515}$ $6\overline{)398}$

3. $5\overline{)467}$ $9\overline{)698}$ $8\overline{)300}$ $2\overline{)199}$ $7\overline{)519}$

4. $4\overline{)270}$ $3\overline{)245}$ $7\overline{)353}$ $9\overline{)431}$ $5\overline{)328}$

If a division problem is written with the ÷ sign, rewrite the problem using the $\overline{)}$ sign. Notice the placement of numbers with these two signs.

Example: 226 ÷ 3 =

Change to: $\begin{array}{r} 75\text{r}3 \\ 3\overline{)226} \end{array}$

B. Rewrite each problem and divide.

5. 1,542 ÷ 8 = 5,050 ÷ 7 = 2,743 ÷ 9 = 1,760 ÷ 3 =

6. 3,845 ÷ 6 = 7,355 ÷ 9= 7,366 ÷ 8 = 2,277 ÷ 5 =

7. 2,183 ÷ 8 = 4,765 ÷ 6= 1,937 ÷ 2 = 2,330 ÷ 4 =

Use the following information to answer questions 8–9.

To make a climbing toy for her children, Mary is sawing pieces of wood each 4 feet long from a piece that is 19 feet long.

8. How many pieces can Mary cut from the long piece?

9. Assuming no waste, what will be the length of the remaining piece?

Use the following information to answer questions 10–11.

Antonio is a carpenter. He needs 9 feet of molding to trim small windows in an attic. He has a total of 40 feet of molding.

10. How many windows can he trim with his supply of molding?

11. If Antonio uses his supply of molding for attic windows, how many feet of molding will be left?

To check your answers, turn to page 176.

Division by Larger Numbers

Sonny borrowed $2,976. He is paying back $62 a month. How many months will he need to repay the entire loan?

To divide by two-digit and three-digit numbers, you must **estimate** how many times one number divides into another number. When you estimate, you make a guess.

Example: Divide 2,976 by 62.

Step 1. **Estimate** how many time 62 divides into 297. To do this, ask yourself how many times 6 divides into 29. $29 \div 6 = 4$ plus a remainder. Write 4 above the 7.

```
     4
62)2,976
```

Step 2. **Multiply:** $4 \times 62 = 248$. Write 248 under 297.

Step 3. **Subtract:** $297 - 248 = 49$, and **compare:** 49 is less than 62.

```
     4
62)2,976
   2 48
     49
```

Step 4. **Bring down the next number:** 6.

Step 5. **Estimate** how many times 62 divides into 496. To do this, ask yourself how many times 6 goes into 49. $49 \div 6 = 8$ plus a remainder. Write 8 above the 6.

Step 6. **Multiply:** $8 \times 62 = 496$. Write 496 under 496.

Step 7. **Subtract:** $496 - 496 = 0$.

Step 8. Check: $48 \times 62 = 2,976$.

```
     48
62)2,976
   2 48
     496
     496
       0
```

Sometimes your first estimate will be wrong. Work with a pencil and a good eraser. Decide if your estimate is too small or too large and try again.

➠ Sonny will need 48 months to repay the loan.

PRACTICE 11

Divide each problem.

1. $24\overline{)192}$ $89\overline{)623}$ $38\overline{)228}$ $62\overline{)310}$

2. $78\overline{)732}$ $18\overline{)105}$ $87\overline{)360}$ $60\overline{)495}$

3. $41\overline{)1{,}968}$ $67\overline{)3{,}752}$ $52\overline{)3{,}380}$ $49\overline{)2{,}548}$

4. $23\overline{)2{,}093}$ $72\overline{)5{,}544}$ $54\overline{)3{,}510}$ $28\overline{)2{,}492}$

Rewrite and divide.

5. $1{,}616 \div 22 =$ $6{,}910 \div 85 =$ $1{,}412 \div 39 =$ $2{,}406 \div 91 =$

6. $2{,}554 \div 86 =$ $2{,}216 \div 44 =$ $6{,}237 \div 76 =$ $2{,}132 \div 29 =$

7. How many two-pound boxes can be filled with 178 pounds of salt?

8. Last year the Melinos paid $7,440 in mortgage payments. There are 12 months in a year. How much did they pay each month?

9. Nora and Doug bought a new TV for $612. They agreed to make 17 equal monthly payments. How much will they pay each month?

10. There are 16 ounces in a pound. How many pounds are there in 560 ounces?

To check your answers, turn to page 177.

Whole Numbers Review

These problems will help you find out if you need to review the whole number section of this book. Solve each problem. When you finish, look at the chart to see which pages you should review.

In problems 1–2, supply the missing words you need to read each number.

1. 60,009,040

sixty ______________, nine ______________, forty

2. 5,300,000

five ______________, three hundred ______________

For problems 3-4, write the numbers in figures.

3. fifteen thousand, two hundred six ____________

4. four million, one hundred twenty thousand, eight ____________

5. What is 328 rounded to the nearest ten?

6. What is 19,512 rounded to the nearest thousand?

7. $\begin{array}{r} 86 \\ +\ 75 \\ \hline \end{array}$

8. $\begin{array}{r} 43 \\ 96 \\ +\ 77 \\ \hline \end{array}$

9. 6,927 + 434 + 56=

10. The distance from Eugene to Portland is 109 miles. The distance from Portland to Seattle is 174 miles. What is the distance from Eugene to Seattle by way of Portland?

11. 7,274 − 5,142 =

12. 800 − 73 =

13. 50,030 − 8,916 =

14. 403 people belong to the Midvale Employees' Union. Of these, 287 voted to strike. How many members did not vote to strike?

15. 73 × 64 =

16. 26 × 785 =

17. 4,086 × 39 =

18. Joelle can type 83 words per minute. How many words can she type in 12 minutes?

19. $52\overline{)4{,}836}$

20. 3,960 ÷ 8 =

21. 3,627 ÷ 42 =

22. Colin can drive 24 miles on one gallon of gasoline. How many gallons will he need to go on a 768-mile trip?

PROGRESS CHECK

Check your answers on page 177. Then turn to the review pages for the problems you missed. Correct your answers before going on to the next unit.

If you missed problems	*Review pages*
1 to 4	7 to 9
5 to 6	10
7 to 10	11 to 15
11 to 14	16 to 19
15 to 18	20 to 21
19 to 22	22 to 27

UNIT

2

Decimals

Place Value

A decimal is a kind of fraction. Like a fraction, a decimal shows a part of a whole. Decimals divide a whole into 10 parts or 100 parts or 1,000 parts and so on. You have used decimal since you first handled money.

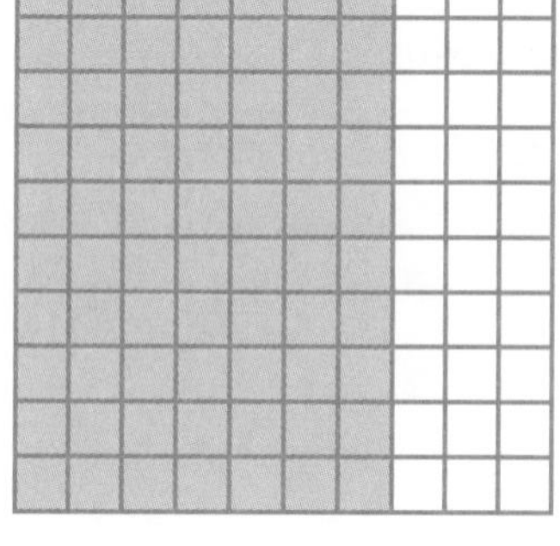

$.70 is a decimal. It is 70 of the 100 equal parts in a dollar. The square pictured here represents a dollar. The shaded part represents 70¢ or 70 of the 100 equal parts of a dollar.

Decimals get their names from the number of **places** on the right side of the decimal point. The decimal point separates whole numbers from decimals. A *place* is the position of a digit. The decimal point itself does not take up a place. The decimal .70 has two places. Learn the place names shown here.

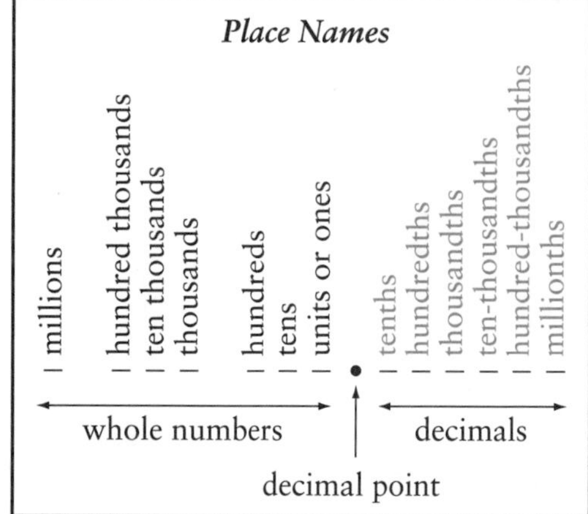

Mixed decimals are numbers with digits on both sides of the decimal point. \$4.95 is a mixed decimal. It means 4 whole dollars and $\frac{95}{100}$ of a dollar. Mixed decimals have whole numbers at the left of the decimal point.

As you move to the right in the decimal system, each place means that the whole has been divided into more parts. As you move to the right, the *values* of the decimal places get smaller. Learn the place values shown in the chart.

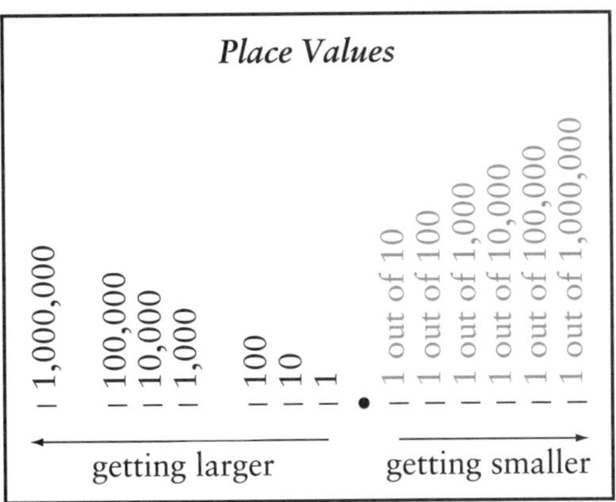

Example: What is the value of 6 in the number .68?

The digit 6 is in the tenths place. It has a value of 6 out of ten equal parts in one whole.

PRACTICE 12

Answer each question.

1. Underline the digit in the tenths place in each number.
   3.2 .189 .02 29.5 .4683

2. Underline the digit in the hundredths place in each number.
   5.38 .279 .46 84.13 3.4921

3. Underline the digit in the thousandths place in each number.
   .345 2.009 .4872 12.1185 .0023

Use the number 5.39 to answer problems 4–7.

4. The digit 3 is in the ________________ place.

5. The digit 3 has a value of 3 out of the ________________ equal parts in one whole.

6. The digit 9 is in the ________________ place.

7. The digit 9 has a value of 9 out of the ________________ equal parts in one whole.

To check your answers, turn to page 177.

Reading and Writing Decimals

Remember that a decimal gets its name from the number of places at the **right** of the decimal point. To read a decimal, count the places at the right of the point.

Example: Read the decimal .042.

Count the decimal places. The decimal .042 has three. Three places are thousandths. Read .042 as *forty-two thousandths.*

Do not confuse a period at the end of a sentence with a decimal point.

With mixed decimals, separate the whole number and the decimal with the word *and.*

Example: Read 13.09.

Count the decimal places. The mixed decimal 13.09 has two. Two decimal places are hundredths. Read 13.09 as *thirteen and nine hundredths.*

PRACTICE 13

A. Supply the missing words you need to read each number.

1. .3 = three ________________

2. .06 = six ________________

3. .015 = fifteen ________________

4. 4.2 = four ________________ two ________________

5. 8.07 = eight ________________ seven ________________

When you write decimals, decide how many places you need. Use zeros in places that are not filled.

Example: Write eleven and nine thousandths as a mixed decimal.

Step 1.	Write the whole number 11.	11
Step 2.	Decide how many places you need. Thousandths need three places.	
Step 3.	The number 9 needs only one place. Put zeros in the first two decimal places.	11.009

B. Write each number as a decimal or a mixed decimal.

6. three tenths ____________

7. thirteen thousandths ____________

8. two hundredths ____________

9. five and four hundredths ____________

10. thirty and seven tenths ____________

11. twelve ten-thousandths ____________

12. sixteen millionths ____________

13. twenty-six and nine tenths ____________

To check your answers, turn to page 177.

Comparing Decimals

Cal wants to know which is heavier, a package that weighs .08 pound or a package that weighs .6 pound.

When you compare decimals, first change the decimals to new decimals with the same number of places. You can put zeros to the right of a decimal without changing its value. For example .5 and .50 have the same value. The 5 is in the tenths place in each decimal.

Example: Which decimal is greater, .08 or .6?

STEP 1. Put a zero at the right of .6 to change it to .60. Both decimals are hundredths now. .60

STEP 2. Decide which is greater .08 or .60. Sixty hundredths is more than eight hundredths. The decimal .6 is the greater. $60 > 8$

The symbol > means "is greater than."

➡ The package that weighs .6 pound is heavier.

Example: Which decimal has the greatest value, .34, .3, or .304?

Step 1.	Put a zero at the right of .34 to change it to .340.	.340
Step 2.	Put two zeros at the right of .3 to change it to .300. All three decimals are thousandths now.	.300
Step 3.	Decide which is biggest, .340, .300, or .304. Three hundred forty thousandths is the biggest. .34 has the greatest value.	

PRACTICE 14

Circle the greater decimal in each pair.

1. .9 or .95	.27 or .3	.07 or .052
2. .297 or .4	.004 or .04	.05 or .061
3. .64 or .626	.33 or .323	.564 or .55

Circle the greatest decimal in each group.

4. .7, .07, or .67	.407, .43, or .4	.0012, .201, or .12
5. .29, .3, or .302	.5, .055, or .505	.707, .77, or .07
6. .08, .028, or .82	.79, .097, or .709	.033, .03, or .3303

To check your answers, turn to page 178.

Rounding Decimals

Sandra shipped a package that weighed 2.48 pounds. Find the weight of the package to the nearest tenth of a pound.

To find the weight of the package to the nearest tenth of a pound, you must *round off* 2.48. Rounding makes numbers easier to use when you don't need exact values. To round off a number, you must know the place value of each digit in the number.

To round a decimal:

1. Underline the digit in the place to which you want to round.
2. If the digit to the right of the underlined digit is more than 4, add 1 to the underlined digit.
3. If the digit to the right of the underlined digit is less than 5, do not change the underlined digit.
4. Drop the digits to the right of the underlined digit.

Example: Round 2.48 to the nearest tenth.

STEP 1.	Underline the digit in the tenths place, 4.	2.48
STEP 2.	The digit to the right of 4 is 8. Add 1 to 4, and drop the digits to the right.	2.5

➡ To the nearest tenth of a pound, the package weighs 2.5 pounds.

Example: Round .732 to the nearest hundredth.

STEP 1.	Underline the digit in the hundredths place, 3.	.732
STEP 2.	The digit to the right of 3 is 2. Do not change 3, but drop the digit to the right.	.73

➡ .732 to the nearest hundredth is .73.

Example: Round .0198 to the nearest thousandth.

STEP 1.	Underline the digit in the thousandths place, 9.	.0198
STEP 2.	The digit to the right of 9 is 8. Add 1 to 9. Since 1 + 9 = 10, you must carry 1 over to the hundredths column. Drop the digit to the right of the thousandths.	.020

➡ .0198 to the nearest thousandth is .020.

PRACTICE 15

Solve.

1. Round each number to the nearest tenth.

.39	2.13	.454	8.276	7.98

2. Round each number to the nearest hundredth.

.073	.635	12.498	.4239	6.337

3. Round each number to the nearest thousandth.

.1047	2.8817	.0066	.0395	4.4892

4. Round each number to the nearest unit or whole number.

7.36	2.552	9.74	307.2	41.83

Use the following information to answer problems 5–6.

One meter is equal to 1.0936 yards.

5. Find the amount to the nearest hundredth yard.

6. Find the amount to the nearest tenth yard.

To check your answers, turn to page 178.

Addition of Decimals

Richard is a plumber. He wants to know the combined thickness of three copper fittings. One is .63 inch thick, the second is 2 inches thick, and the third is 1.279 inches thick.

To add decimals, line up the numbers with the decimal points under each other.

Example: .63 + 2 + 1.279 =

STEP 1. Line up the numbers with the points under each other.

STEP 2. Add each column.

$$\begin{array}{r} .63 \\ 2. \\ +\ 1.279 \\ \hline 3.909 \end{array}$$

Remember that a whole number is understood to have a decimal point to the right.

➡ The combined thickness of the fittings is 3.909 inches.

PRACTICE 16

Add each problem.

1. .28 + .3 + .709 = .34 + .959 + .6 =

2. .3 + .8 + .6 = .27 + .94 + .08 =

3. .68 + .7 + .697 = .3 + .4177 + .274 =

4. 2.1 + 66 + 3.97 = .506 + 5.6 + 4 =

5. 70 + 6.256 + .49 = 5 + .92 + .747 =

6. The average April temperature in Chicago is 47.8°. The average April temperature in St. Louis is 8.3° higher than in Chicago. Find the average April temperature for St. Louis.

7. In 1995 the town of Elmford spent $2.2 million for education. In 1996 it spent $.85 million more. How much did it spend on education in 1996?

8. On Monday Ann drove 3.7 miles to take her children to school. She also drove 1.9 miles to a gas station, 4.6 miles to a shopping center and 5.8 miles back home. How many miles did she drive in total?

9. In 1970 there were about 3.7 billion people in the world. In the year 2000 there will be about .8 billion more people. What will be the world population in 2000?

10. The reading on the mileage gauge of Pete's car was 36,405.2 miles on Friday morning. By Sunday night Pete had driven 768.9 more miles. What was the reading Sunday night?

11. Rachel's normal temperature is 98.6°. When she was ill, her temperature went up 4.9°. What was her temperature when she was ill?

12. Jack works part-time at a garage. Monday he worked 4.5 hours. Wednesday he worked 3.25 hours. Friday he worked 5 hours. How many hours did he work altogether that week?

13. At Software Depot, Nita bought a computer game for $35.96, a hint book for $9.49, and a joystick for $39.95. How much did she spend before tax?

To check your answers, turn to page 178.

Subtraction of Decimals

From a board 5 meters long, Jamal cut a piece .38 meter long. How long was the remaining piece?

To subtract decimals line up the decimals with the points under each other just like addition. Remember to put a point at the right of a whole number. Put zeros at the right until each decimal has the same number of places. You will need the zeros for borrowing.

Example: Subtract 5 − .38 =

STEP 1. Put a decimal point at the right of 5.

$$\begin{array}{r} 5. \\ -\ .38 \\ \hline \end{array}$$

STEP 2. Line up the numbers with the points under each other.

STEP 3. Put two zeros at the right of 5 to give each decimal the same number of places.

STEP 4. Regroup and subtract.

$$\begin{array}{r} \overset{4}{\cancel{5}}.\overset{\overset{9}{\cancel{10}}}{\cancel{0}}\overset{10}{\cancel{0}} \\ -\ .38 \\ \hline 4.62 \end{array}$$

You can also show regrouping like this:

$$\begin{array}{r} \overset{4}{\cancel{5}}.\overset{9}{\cancel{0}}{}^{1}0 \\ -\ .38 \\ \hline 4.62 \end{array}$$

➡ The remaining piece was 4.62 meters long.

PRACTICE 17

Subtract each problem.

1. 6 − .359 =   12 − .35 =   .7 − .482 =

2. .38 − .098 =   4 − .059 =   .02 − .004 =

3. 3.8 − 2.947 =   5.8 − .399 =   .63 − .406 =

4. 20 − 3.89 =   6 − 2.075 =   8.76 − 3 =

5. .07 − .052 = 30 − .8 = .3 − .049 =

6. 3 − .35 = 12.9 − 10.06 = .936 − .08 =

7. The area of the United States is about 3.3 million square miles. The area of Canada is about 3.8 million square miles. How much bigger is Canada in area?

8. There are about 257.9 million people living in the U.S. About 27.8 million people live in Canada. How many more people live in the U.S. than in Canada?

9. When George bought his used car, the mileage gauge read 15,023.4 miles. In two months the gauge read 19,376.8 miles. How many miles did George drive the first two months?

10. Ty Cobb's batting average for his career was .367. Rogers Hornsby's average was .358. How much better was Ty Cobb's average?

11. Judith is 1.6 meters tall. Her daughter Emma is 1.35 meters tall. How much taller is Judith than her daughter?

12. In 1980 there were 7.5 million people living in the Chicago area. In 1990 there were 8.1 million in that area. By how much did the population grow from 1980 to 1990?

13. Deion bought a piece of lumber 2 meters long. From it he cut a piece 1.85 meters long. How long was the leftover piece?

To check your answers, turn to page 179.

Multiplication of Decimals

Sean has to ship a package that weighs 6.5 pounds to a foreign country. In order to fill out a customs form, he needs to know the weight of the package in kilograms. One pound is .45 kilograms.

To multiply decimals, count the decimal places in each number. Put the total number of decimal places in the answer.

Example: .45 × 6.5 =

Step 1. Multiply the numbers.

Step 2. Count the decimal places in each number: .45 has two decimal places and 6.5 has one.

Step 3. Put the total number of decimal places (2 + 1 = 3) in the answer.

```
   6.5   one decimal place
 × .45   two decimal places
 -----
  32 5
 2 60
 -----
 2.92 5  three decimal places
```

➠ The package weighs 2.925 kilograms.

PRACTICE 18

A. Multiply each problem.

1. .6 × 9.1 =	.8 × 7.23 =	.974 × 7=
2. .4 × 83.9 =	1.9 × 8.6 =	.96 × 3.3 =
3. 7 × 2.6 =	.82 × 9 =	8 × .03 =
4. .23 × 71 =	348 × .05 =	.16 × 352 =
5. .12 × 3.5 =	3.6 × 1.8 =	25 × 5.25 =

Sometimes you will need to put extra zeros in your answer.

Example: .3 × .07 =

STEP 1. Multiply the numbers.

STEP 2. Count the decimal places: .07 has two and .3 has one.

STEP 3. Put the total number of places (2 + 1 = 3) in the answer. Put a zero to the left of 21 to make three places.

.07	two decimal places
× .3	one decimal place
.021	three decimal places

B. Multiply each problem.

6. .09 × .8 =   .047 × .4 =   .006 × .07 =

7. .048 × .66 =   .0635 × 4 =   1.2 × .009 =

8. Jose weighs 180 pounds. One pound equals .45 kilograms. What is Jose's weight in kilograms?

9. Adrienne works overtime for $18.60 an hour. Last week she worked 7.5 hours overtime. How much did she make for overtime work?

10. Ruby bought 2.3 pounds of chicken at $1.79 a pound. How much did she pay? Round your answer to the nearest cent.

11. Fred walks at an average speed of 3.6 miles per hour. How far can he walk in 2.5 hours?

12. Mark bought 4.25 feet of lumber. The lumber cost $3.40 a foot. What was the total cost of the lumber?

To check your answers, turn to page 179.

Division of Decimals by Whole Numbers

Larry wants to cut a board that is 4.68 meters long into six equal pieces to make shelves. Assuming there is no waste to the cuts, how long will each shelf be?

To divide a decimal by a whole number, line up the problem carefully. Then divide and bring the decimal point up into the answer above its position in the problem.

Example: 4.68 ÷ 6 =

STEP 1. Rewrite the problem and divide.

STEP 2. Bring the decimal point up into the answer above its position in the problem.

```
   .78
6)4.68
  4 2
    48
    48
     0
```

The decimal points in the problem and the answer line up.

➠ Each shelf will be .78 meter long.

PRACTICE 19

A. Divide each problem.

1. 29.6 ÷ 8 = 3.12 ÷ 6 = 13.16 ÷ 4 =

2. 4.368 ÷ 7 = 2.01 ÷ 3 = 67.2 ÷ 24 =

Sometimes you will need to put zeros in your answers.

Example: .512 ÷ 8 =

STEP 1. Rewrite the problem and divide.

STEP 2. Bring the decimal point up into the answer above its position in the problem.

STEP 3. To show that 8 does not divide into .5, put a zero above the 5.

```
  .064
8).512
   42
    32
    32
     0
```

B. Divide each problem.

3. $.342 \div 9 =$ $.324 \div 12 =$ $.702 \div 18 =$

4. $1.633 \div 23 =$ $2.88 \div 32 =$ $.136 \div 8 =$

5. $\$1.48 \div 4 =$ $\$.96 \div 12 =$ $\$4.20 \div 3 =$

6. $\$8.25 \div 11 =$ $\$1.35 \div 45 =$ $\$10.08 \div 9 =$

7. Jake is a plumber. He wants to cut a piece of pipe 2.52 meters long into four equal pieces. How long will each piece be?

8. John works 35 hours a week. In a week he makes $505.75 before taxes. How much does John make in one hour?

To check your answers, turn to page 179.

Division of Decimals by Decimals

How many pieces of pipe, .8 meter long each, can Tony cut from a pipe that is 3.68 meters long?

To divide a decimal by a decimal, first make a new problem. Change the number you are dividing by (the *divisor*) into a whole number. You can change the divisor into a whole number by moving the decimal point to the right end. Then move the decimal point in the other number (the *dividend*) the same number of places.

These steps are easier to understand with whole numbers. Think about the problem $10 \div 2 = 5$. The answer is the same if the decimal point moves to the right in both 10 and 2. $100 \div 20 = 5$. The problems are different, but the answers are the same.

$$2\overline{)10} = 5$$

$$2.0\overline{)10.0} = 5$$

Example: $3.68 \div .8 =$

STEP 1. Rewrite the problem, and move the decimal point in the divisor, .8, one place to the right to make it a whole number.

$.8\overline{)3.68}$

STEP 2. Move the decimal point in the dividend, 3.68, one place to the right.

$.8\overline{)3.6\,8}$

STEP 3. Divide, and bring the decimal point up into the answer above its new position.

$$\begin{array}{r} 4.6 \\ .8\overline{)3.6\,8} \\ \underline{3\,2} \\ 4\,8 \\ \underline{4\,8} \\ 0 \end{array}$$

Bring the decimal point up.

Remember to move the decimal point in the divisor and the decimal point in the dividend the same *number of places.*

➠ Tony can cut 4.6 or 4 complete pieces.

PRACTICE 20

A. Divide each problem.

1. $2.38 \div .7 =$ $.504 \div .9 =$ $2.34 \div .3 =$

2. $.567 \div .09 =$ $.0348 \div .06 =$ $.072 \div .04 =$

Sometimes you will have to put extra zeros in the dividend.

Example: $5.6 \div .07 =$

STEP 1. Rewrite the problem, and move the decimal point in the divisor, .07, two places to the right to make it a whole number.

$.07\overline{)5.6}$

STEP 2. Move the decimal point in the dividend, 5.6, two places to the right. Put an extra zero to the right of 5.6 to get two decimal places.

$.07\overline{)5.60}$

STEP 3. Divide. Since the answer is a whole number, no decimal point is needed.

$$\begin{array}{r} 80 \\ .07\overline{)5.60} \\ \underline{5\,6} \\ 00 \end{array}$$

B. Divide each problem.

3. 7.8 ÷ .003 = 5.31 ÷ .009 = 85.02 ÷ .026 =

4. 2.82 ÷ .006 = 55.2 ÷ .92 = 40.45 ÷ .809 =

5. Charlene drove 267.5 miles on 12.5 gallons of gas. How far did she drive on one gallon?

To check your answers, turn to page 180.

Division of Whole Numbers by Decimals

How many nickels ($.05) are there in eight dollars ($8)?

To divide a whole number by a decimal, remember to put a decimal point at the right of the whole number. Then move the points in both the divisor and the dividend. You will have to put zeros in the dividend.

Example: 8 ÷ .05 =

Step 1. Rewrite the problem, and move the decimal point in the divisor, .05, two places to the right to make it a whole number.

.05)8

Step 2. To move the decimal point in the dividend, 8, first put a decimal point to the right of 8, then put two zeros to the right and move the point.

.05)8.00

Step 3. Divide. Since there is no decimal part to the answer, you can drop the decimal point.

1 60
.05)8.00

There are 160 nickels in $8.

PRACTICE 21

A. Divide each problem.

1. $36 \div 2.4 =$ $\quad$ $30 \div .75 =$ $\quad$ $39 \div .06 =$

2. $54 \div 4.5 =$ $\quad$ $20 \div .08 =$ $\quad$ $18 \div .036 =$

Not every division problem comes out even. When this happens, choose a place to round to. Then divide one place beyond the place you want to round to.

To review rounding, turn to page 10.

Example: Find $8 \div 1.5$ to the nearest tenth.

STEP 1. Since you want an answer to the nearest tenth, divide to the hundredths place. Put two zeros to the right of the new position of the decimal point. Then divide.

```
      5.33
1.5)8.0 00
    7 5
    ---
      5 0
      4 5
      ---
       50
       45
```

STEP 2. Round the answer to the nearest tenth. 5.33 to the nearest tenth is 5.3

B. For row 3, round each answer to the nearest tenth.

3. $9 \div 1.3 =$ $\quad$ $12 \div .7 =$ $\quad$ $5 \div .3 =$

For row 4, round each answer to the nearest hundredth.

4. $20 \div 1.4 =$ $\quad$ $18 \div 3.1 =$ $\quad$ $1 \div .85 =$

5. A 50-acre parcel of land is to be divided up into three equal pieces. Find, to the nearest tenth, the number of acres in each piece.

6. Sandy paid \$6 for 2.2 pounds of lamb. To the nearest cent, what was the price of one pound of lamb?

To check your answers, turn to page 180.

Decimals Review

These problems will help you find out if you need to review the decimals section of this book. When you finish, look at the chart to see which pages you should review.

Solve each problem.

For problems 1-2, write each decimal or mixed decimal in words.

1. .004 ______________________________

2. 8.1 ______________________________

For problems 3–4, write each number as a decimal or a mixed decimal.

3. eighteen thousandths ____________

4. one and eight hundredths __________

5. Which decimal is greater, .52 or .504?

6. Round 2.38 to the nearest tenth.

7. .0052 + .84 + .072 =

8. .26 + 14.7 + 13 =

9. According to the 1970 census, there were 8.4 million people in the Los Angeles area. In 1990 there were 6.1 million more people. How many people lived in Los Angeles in 1990?

10. 11 − .509 = *11.* 8.3 − 2.052 = *12.* 3.24 − .966 =

13. Jorge is 1.9 meters tall. His son Mateo is 1.05 meters tall. How much taller is Jorge than Mateo?

14. .47 × 9 = *15.* 3.4 × 1.9 = *16.* 4.3 × .38 =

17. Jennifer makes $14.90 an hour. How much does she make on a day when she works 8.5 hours?

18. 110 ÷ 23 = *19.* .621 ÷ .09 = *20.* 54 ÷ .27 =

21. Joe wants to cut a board that is 10.8 feet long into four equal pieces. How long will each piece be?

PROGRESS CHECK

Check your answers on page 180. Then return to the review pages for the problems you missed. Correct your answers before going on to the next unit.

If you missed problems	*Review pages*
1 to 4	32 to 35
5 to 6	36 to 38
7 to 9	39 to 40
10 to 13	41 to 42
14 to 17	43 to 44
18 to 21	45 to 50

UNIT 3

Fractions

Writing Fractions

A fraction is made up of two numbers that show a part of some whole. For example, seven dimes are seven of the ten equal parts of a dollar. Seven dimes are $\frac{7}{10}$ or seven tenths of a dollar.

The top number of a fraction is called the **numerator.** The numerator tells how many parts you have. The bottom number is called the **denominator.** The denominator tells how many parts are in the whole.

In the fraction $\frac{7}{10}$, the number 7 is the numerator, and 10 is the denominator. You have 7 parts. The whole has 10 parts.

The pictures below are partly shaded. The fraction next to each picture tells what part of the picture is shaded. The numerator tells how many parts are shaded. The denominator tells how many parts are in the whole picture.

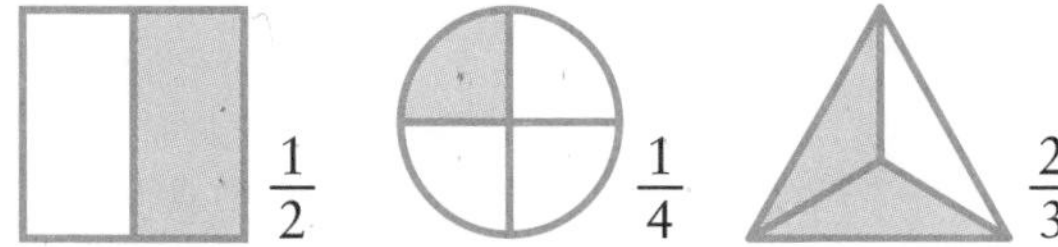

PRACTICE 22

A. For each picture write a fraction that tells what part of the picture is shaded.

1. 

2. 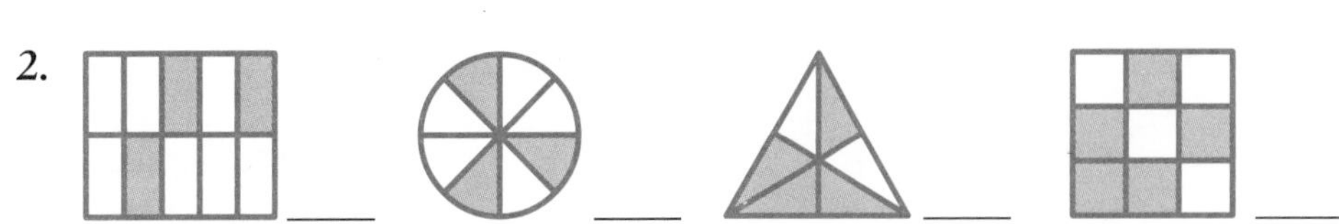

To write a fraction from words, first choose a denominator that tells the number of parts in the whole. Then choose a numerator that tells how many parts you have.

Example: There are 12 inches in a foot. 11 inches are what fraction of a foot?

STEP 1. Choose the denominator. 12 tells how many parts are in one whole foot. 12 is the denominator. $\frac{\ }{12}$

STEP 2. Choose the numerator. 11 tells how many parts you have. 11 is the numerator. $\frac{11}{12}$

➡ Eleven inches are $\frac{11}{12}$ of a foot.

B. Write a fraction for each problem.

3. There are 4 quarts in a gallon. Three quarts are what fraction of a gallon?

4. There are 10 dimes in a dollar. Nine dimes are what fraction of a dollar?

5. There are 60 minutes in an hour. 21 minutes are what fraction of an hour?

6. There are 1000 grams in a kilogram. 127 grams are what fraction of a kilogram?

7. There are 36 inches in a yard. 19 inches are what fraction of a yard?

8. There are 7 days in a week. Two days are what fraction of a week?

To check your answers, turn to page 181.

Identifying Forms of Fractions

There are three forms of fractions.

Proper fraction: The numerator (top number) is always less than the denominator. Example: $\frac{2}{3}$

The value of a proper fraction is less than one whole.

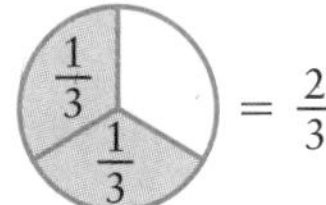

Improper fraction: The numerator is equal to or greater than the denominator. Examples: $\frac{4}{4}$ and $\frac{4}{3}$

When the numerator is equal to the denominator, an improper fraction is equal to one whole.

When the numerator is greater than the denominator, an improper fraction has a value of more than one whole.

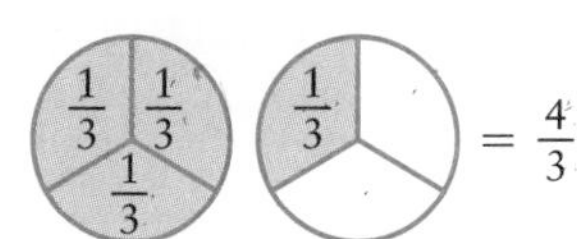

Mixed number: A whole number and a proper fraction are written next to each other. Example: $1\frac{1}{2}$

A mixed number always has a value of more than one whole.

PRACTICE 23

Answer each question.

1. Circle the proper fractions. $\frac{6}{7}$ $\frac{8}{3}$ $\frac{4}{5}$ $\frac{5}{4}$ $\frac{2}{2}$ $9\frac{5}{7}$ $\frac{8}{200}$

2. Circle the improper fractions. $\frac{19}{5}$ $\frac{5}{12}$ $8\frac{5}{6}$ $\frac{12}{9}$ $4\frac{2}{5}$ $\frac{15}{15}$

3. Circle the mixed numbers. $\frac{9}{15}$ $8\frac{4}{7}$ $\frac{16}{15}$ $2\frac{3}{20}$ $\frac{19}{24}$ $3\frac{8}{9}$

To check your answers, turn to page 181.

Reducing

In the first circle pictured here, two fourths are shaded. In the second circle, one half is shaded. The same part of each circle is shaded. The fraction $\frac{2}{4}$ *reduces* to $\frac{1}{2}$.

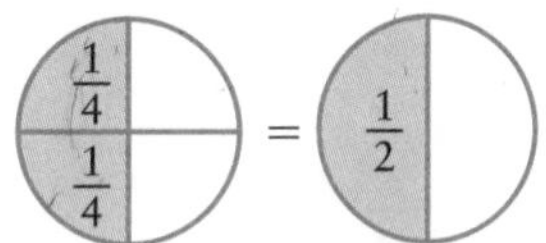

Reducing a fraction means dividing both the numerator and the denominator (the top and the bottom) by a number that goes into each evenly. Reducing changes the numbers in a fraction, but reducing does not change the value of a fraction. Remember that $\frac{1}{2}$ dollar or a 50¢ piece is the same amount of money as $\frac{2}{4}$ dollar or two 25¢ pieces.

Example: Reduce $\frac{18}{20}$.

STEP 1. Divide both 18 and 20 by a number that goes evenly into both of them. 2 divides evenly into both 18 and 20.

$$\frac{18 \div 2}{20 \div 2} = \frac{9}{10}$$

STEP 2. Check to see if another number, besides 1, goes evenly into 9 and 10. No other number divides evenly into both. The fraction $\frac{9}{10}$ is reduced as far as it can be.

$$\frac{18}{20} = \frac{9}{10}$$

The equal sign (=) tells you that $\frac{9}{10}$ has exactly the same value as $\frac{18}{20}$.

Sometimes a fraction can be reduced more than once.

Example: Reduce $\frac{32}{48}$.

STEP 1. Divide both 32 and 48 by a number that goes evenly into both of them. 8 divides evenly into both 32 and 48. $\frac{32 \div 8}{48 \div 8} = \frac{4}{6}$

STEP 2. Check to see if another number goes evenly into 4 and 6. The number 2 divides evenly into both numbers. Divide 4 and 6 by 2. $\frac{4 \div 2}{6 \div 2} = \frac{2}{3}$

STEP 3. Check to see if another number goes evenly into 2 and 3. No other number divides evenly into both. $\frac{2}{3}$ is reduced as far as it can be. $\frac{32}{48} = \frac{4}{6} = \frac{2}{3}$

A fraction that is reduced as far as it can be is in **lowest terms.**

PRACTICE 24

Reduce each fraction to lowest terms.

1. $\frac{35}{42} =$ $\frac{32}{40} =$ $\frac{20}{32} =$ $\frac{18}{45} =$ $\frac{12}{30} =$

2. $\frac{9}{33} =$ $\frac{22}{24} =$ $\frac{36}{48} =$ $\frac{25}{45} =$ $\frac{56}{84} =$

3. $\frac{36}{63} =$ $\frac{24}{40} =$ $\frac{15}{27} =$ $\frac{13}{26} =$ $\frac{8}{96} =$

4. $\frac{16}{64} =$ $\frac{5}{105} =$ $\frac{70}{110} =$ $\frac{28}{32} =$ $\frac{49}{63} =$

Write a fraction for each problem. Reduce each fraction to lowest terms.

5. Mr. and Mrs. Smith make $1,840 a month. They spend $460 a month for rent. What fraction of their income goes for rent?

6. Jason took a test with 60 problems. He got 48 problems right. What fraction of the problems did he get right?

7. Shirley weighed 140 pounds. She went on a diet and lost 20 pounds. What fraction of her weight did she lose?

To check your answers, turn to page 181.

Raising Fractions to Higher Terms

In the picture here, $\frac{3}{4}$ of a circle is shaded. This is the same as $\frac{6}{8}$.

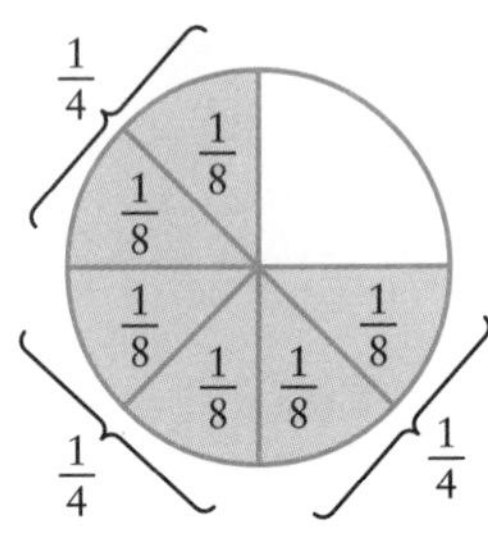

Raising to higher terms is the opposite of reducing. To raise a fraction to higher terms, multiply both the numerator and the denominator by the same number.

Example: Raise the fraction to higher terms by finding the missing numerator. $\frac{3}{4} = \frac{\quad}{8}$

STEP 1. Divide the new denominator by the original denominator. $4\overline{)8}$ = 2

STEP 2. Multiply the original numerator and the original denominator by 2. The fraction $\frac{6}{8}$ is $\frac{3}{4}$ raised to higher terms. $\frac{3 \times 2}{4 \times 2} = \frac{6}{8}$

➡ $\frac{3}{4}$ is equal to $\frac{6}{8}$.

PRACTICE 25

Raise each fraction to higher terms by finding the missing numerator.

1. $\frac{3}{4} = \frac{\quad}{24}$ $\frac{2}{9} = \frac{\quad}{36}$ $\frac{7}{10} = \frac{\quad}{80}$ $\frac{2}{5} = \frac{\quad}{35}$ $\frac{5}{8} = \frac{\quad}{40}$

2. $\frac{6}{7} = \frac{\quad}{63}$ $\frac{1}{12} = \frac{\quad}{36}$ $\frac{3}{5} = \frac{\quad}{50}$ $\frac{4}{9} = \frac{\quad}{54}$ $\frac{8}{11} = \frac{\quad}{22}$

3. $\frac{2}{3} = \frac{\quad}{36}$ $\frac{2}{9} = \frac{\quad}{72}$ $\frac{7}{10} = \frac{\quad}{60}$ $\frac{1}{8} = \frac{\quad}{56}$ $\frac{4}{5} = \frac{\quad}{45}$

To check your answers, turn to page 181.

Changing Improper Fractions to Whole or Mixed Numbers

In the circles pictured here, a total of $\frac{6}{4}$ are shaded. This is the same as $1\frac{2}{4}$ or $1\frac{1}{2}$.

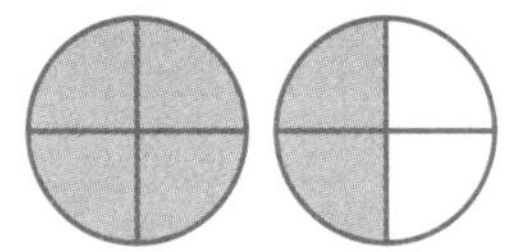

The answers to many fraction problems are improper fractions. These answers are easier to read if you change them to whole numbers or mixed numbers. In an improper fraction the numerator (top number) is as big or bigger than the denominator (bottom number). To change an improper fraction, divide the denominator into the numerator.

Example: Change $\frac{6}{4}$ to a mixed number.

STEP 1. Divide the denominator into the numerator.

$$\begin{array}{r} 1 \\ 4\overline{)6} \\ \underline{4} \\ 2 \end{array}$$

STEP 2. Write the answer to the division as the whole number. Write the remainder over the old denominator.

$$\frac{6}{4} = 1\frac{2}{4}$$

STEP 3. Reduce $\frac{2}{4}$ by 2.

$$1\frac{2 \div 2}{4 \div 2} = 1\frac{1}{2}$$

➡ $\frac{6}{4}$ is equal to $1\frac{1}{2}$.

PRACTICE 26

Change each improper fraction to a whole number or a mixed number. Reduce each fraction that is left.

1. $\frac{5}{2} =$ $\quad\frac{13}{4} =$ $\quad\frac{17}{3} =$ $\quad\frac{43}{5} =$ $\quad\frac{40}{9} =$

2. $\frac{37}{10} =$ $\quad\frac{24}{4} =$ $\quad\frac{18}{7} =$ $\quad\frac{53}{9} =$ $\quad\frac{56}{7} =$

3. $\frac{22}{12} =$ $\quad\frac{41}{12} =$ $\quad\frac{30}{6} =$ $\quad\frac{22}{8} =$ $\quad\frac{23}{7} =$

To check your answers, turn to page 182.

Changing Mixed Numbers to Improper Fractions

In the circles pictured here, a total of $2\frac{1}{2}$ are shaded. This is the same as $\frac{5}{2}$.

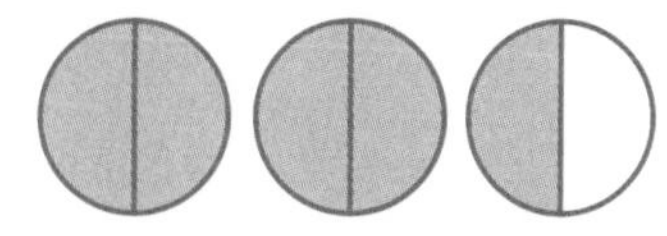

When you multiply and divide fractions, you will have to change mixed numbers to improper fractions. To change a mixed number to an improper fraction, follow these steps:

1. Multiply the denominator (bottom number) by the whole number.
2. Add the numerator.
3. Write the total over the denominator.

Example: Change $2\frac{1}{2}$ to an improper fraction.

STEP 1. Multiply the denominator, 2, by the whole number, 2: $2 \times 2 = 4$. $\quad 2\frac{1}{2} = \frac{}{2}$

STEP 2. Add the numerator: $4 + 1 = 5$. $\quad 2\frac{1}{2} = \frac{5}{2}$

STEP 3. Write the total, 5, over the denominator, 2: $\frac{5}{2}$.

➡ $2\frac{1}{2}$ is equal to $\frac{5}{2}$.

PRACTICE 27

Change each mixed number to an improper fraction.

1. $6\frac{2}{3} =$ $\quad 3\frac{1}{2} =$ $\quad 5\frac{3}{4} =$ $\quad 2\frac{7}{10} =$ $\quad 1\frac{1}{6} =$

2. $3\frac{5}{8} =$ $\quad 9\frac{2}{3} =$ $\quad 4\frac{2}{7} =$ $\quad 6\frac{5}{6} =$ $\quad 1\frac{9}{10} =$

3. $8\frac{1}{3} =$ $\quad 2\frac{5}{12} =$ $\quad 9\frac{1}{4} =$ $\quad 4\frac{7}{8} =$ $\quad 3\frac{6}{7} =$

To check your answers, turn to page 182.

Comparing Fractions

To compare the size of two proper fractions, decide which fraction is closer in value to one whole. It is sometimes hard to compare the size of two fractions when both the numerators and the denominators are different. When the numerators and denominators are different, find a **common denominator** for both fractions. A common denominator is a number that both denominators divide into evenly.

Example: Which fraction is greater, $\frac{3}{5}$ or $\frac{5}{7}$?

STEP 1. Find a common denominator for 5 and 7. The new denominator should be the lowest number that both 5 and 7 divide into evenly. Both 5 and 7 divide into 35 evenly.

$7 \times 5 = 35$

To find a common denominator, keep multiplying the larger denominator until you find a number that both denominators will go into evenly.

$7 \times 1 = 7$
$7 \times 2 = 14$
$7 \times 3 = 21$
$7 \times 4 = 28$
$7 \times 5 = \underline{35}$

STEP 2. Raise each fraction to a new fraction with a denominator of 35.

$$\frac{3 \times 7}{5 \times 7} = \frac{21}{35}$$
$$\frac{5 \times 5}{7 \times 5} = \frac{25}{35}$$

STEP 3. Decide which fraction is greater. $\frac{25}{35}$ has 4 more 35ths than $\frac{21}{35}$. $\frac{25}{35}$ is bigger. $\frac{25}{35}$ is equal to $\frac{5}{7}$. $\frac{5}{7}$ is the greater fraction.

$$\frac{5}{7} > \frac{3}{5}$$

Remember, the symbol > means "is greater than."

Example: Which fraction is greater, $\frac{2}{5}$ or $\frac{7}{20}$?

STEP 1. Find a common denominator for 5 and 20. The lowest number that both 5 and 20 divide into evenly is 20.

Since 5 divides evenly into 20, you know immediately that 20 is the lowest common denominator.

STEP 2. Raise $\frac{2}{5}$ to a new fraction with a denominator of 20.

$$\frac{2 \times 4}{5 \times 4} = \frac{8}{20}$$

STEP 3. Decide which fraction is greater. $\frac{8}{20}$ has 1 more 20th than $\frac{7}{20}$. $\frac{8}{20}$ is bigger. $\frac{8}{20}$ is equal to $\frac{2}{5}$. $\frac{2}{5}$ is the greater fraction.

$$\frac{2}{5} > \frac{7}{20}$$

PRACTICE 28

A. Circle the greater fraction in each pair.

1. $\frac{1}{2}$ or $\frac{3}{5}$ $\quad$ $\frac{5}{12}$ or $\frac{1}{3}$ $\quad$ $\frac{1}{2}$ or $\frac{8}{15}$ $\quad$ $\frac{7}{9}$ or $\frac{3}{4}$

2. $\frac{4}{5}$ or $\frac{21}{25}$ $\quad$ $\frac{4}{7}$ or $\frac{1}{2}$ $\quad$ $\frac{1}{4}$ or $\frac{1}{6}$ $\quad$ $\frac{8}{15}$ or $\frac{2}{3}$

3. $\frac{2}{5}$ or $\frac{3}{8}$ $\quad$ $\frac{1}{3}$ or $\frac{5}{9}$ $\quad$ $\frac{7}{10}$ or $\frac{3}{4}$ $\quad$ $\frac{5}{16}$ or $\frac{3}{8}$

To compare three fractions, find a common denominator for all three.

Example: Which fraction has the greatest value, $\frac{7}{12}$, $\frac{2}{3}$, or $\frac{5}{8}$?

STEP 1. Find a common denominator for 12, 3, and 8. The lowest number that 12, 3, and 8 divide into evenly is 24.

STEP 2. Raise each fraction to a new fraction with 24 as the denominator.

$$\frac{7 \times 2}{12 \times 2} = \frac{14}{24} \qquad \frac{2 \times 8}{3 \times 8} = \frac{16}{24} \qquad \frac{5 \times 3}{8 \times 3} = \frac{15}{24}$$

STEP 3. Decide which fraction is greatest. $\quad \frac{2}{3} > \frac{5}{8} > \frac{7}{12}$

$\frac{16}{24}$ has 1 more 24th than $\frac{15}{24}$
and it has 2 more 24ths than $\frac{14}{24}$.
$\frac{16}{24}$ is biggest. $\frac{16}{24}$ is equal to $\frac{2}{3}$.
$\frac{2}{3}$ is the greatest fraction.

B. Circle the fraction with the greatest value in each group.

4. $\frac{1}{4}, \frac{3}{8}$, or $\frac{5}{16}$ $\quad$ $\frac{5}{12}, \frac{1}{4}$, or $\frac{5}{24}$ $\quad$ $\frac{3}{5}, \frac{1}{2}$, or $\frac{3}{4}$

5. $\frac{3}{4}, \frac{7}{9}$, or $\frac{29}{36}$ $\quad$ $\frac{9}{20}, \frac{3}{10}$, or $\frac{2}{5}$ $\quad$ $\frac{5}{9}, \frac{2}{3}$, or $\frac{11}{18}$

To check your answers, turn to page 182.

Addition of Fractions with the Same Denominators

The Adams family's house is on a $2\frac{3}{10}$ acre lot. Across the street they own a lot that measures $1\frac{9}{10}$ acres. What is the total measurement of their property?

To find the *total,* add $2\frac{3}{10}$ and $1\frac{9}{10}$. To add fractions with the same denominators (bottom numbers), first add the numerators. Then write the total over the denominator.

Example: $2\frac{3}{10} + 1\frac{9}{10} =$

STEP 1. Add the numerators: 3 + 9 = 12.

STEP 2. Write the total, 12, over the denominator.

STEP 3. Add the whole numbers: 2 + 1 = 3.

STEP 4. Change $\frac{12}{10}$ to a mixed number, $1\frac{2}{10}$.

STEP 5. Add the 1 from $1\frac{2}{10}$ to 3.

STEP 6. Reduce.

$$\begin{array}{r} 2\frac{3}{10} \\ +\,1\frac{9}{10} \\ \hline 3\frac{12}{10} \end{array} = 3 + 1\frac{2}{10} = 4\frac{2}{10} = 4\frac{1}{5}$$

➡ The total acreage of the Adams' property is $4\frac{1}{5}$ acres.

PRACTICE 29

Add and reduce.

1. $\frac{5}{12} + \frac{1}{12} =$ $\qquad 5\frac{1}{4} + 2\frac{1}{4} =$ $\qquad 8\frac{2}{10} + 3\frac{3}{10} =$

2. $\frac{3}{8} + \frac{3}{8} =$ $\qquad 3\frac{13}{24} + 9\frac{7}{24} =$ $\qquad 8\frac{5}{16} + 5\frac{7}{16} =$

3. $8\frac{3}{4} + 6\frac{3}{4} =$ $\qquad 3\frac{7}{9} + 3\frac{5}{9} =$ $\qquad 1\frac{17}{20} + 9\frac{9}{20} =$

To check your answers, turn to page 182.

Addition of Fractions with Different Denominators

The Yoshimuras spend $\frac{5}{12}$ of their income for rent and $\frac{1}{3}$ of their income for food. Together, rent and food use up what fraction of their income?

If the fractions in an addition problem do not have the same denominators, you must find a *common denominator.* A common denominator is a number that can be divided evenly by every denominator in the problem. The lowest denominator that can be divided evenly by every denominator in the problem is called the **lowest common denominator or LCD.**

Example: $\frac{5}{12} + \frac{1}{3} =$

STEP 1. Find the LCD. 3 divides evenly into 12. 12 is the LCD.

STEP 2. Raise $\frac{1}{3}$ to a fraction with 12 as the denominator.

STEP 3. Add the new fractions.

STEP 4. Reduce.

$$\begin{array}{rl} \frac{5}{12} = & \frac{5}{12} \\ + \frac{1}{3} = & \frac{4}{12} \\ \hline & \frac{9}{12} = \frac{3}{4} \end{array}$$

To review raising fractions to higher terms turn to page 56.

➠ The Yoshimuras spend $\frac{3}{4}$ of their income on rent and food.

PRACTICE 30

For each problem, find the lowest common denominator, add, and reduce.

1. $\frac{2}{3} + \frac{5}{6} =$ $\qquad \frac{4}{9} + \frac{7}{8} =$ $\qquad \frac{7}{8} + \frac{3}{4} =$ $\qquad \frac{3}{10} + \frac{4}{5} =$

2. $\frac{3}{4} + \frac{2}{5} =$ $\qquad \frac{2}{3} + \frac{5}{8} =$ $\qquad \frac{3}{5} + \frac{5}{6} =$ $\qquad \frac{1}{2} + \frac{4}{7} =$

3. $\frac{4}{9} + \frac{5}{6} =$ $\qquad \frac{5}{12} + \frac{3}{8} =$ $\qquad \frac{8}{15} + \frac{7}{10} =$ $\qquad \frac{2}{9} + \frac{11}{12} =$

4. $\frac{5}{12} + \frac{3}{8} + \frac{1}{6} =$ $\qquad$ $\frac{1}{2} + \frac{3}{4} + \frac{3}{10} =$ $\qquad$ $\frac{7}{20} + \frac{1}{2} + \frac{5}{8} =$

5. $7\frac{5}{8} + 3\frac{1}{4} + 4\frac{7}{12} =$ $\qquad$ $9\frac{3}{4} + 6\frac{3}{8} + 2\frac{3}{10} =$

6. $5\frac{5}{6} + 2\frac{1}{4} + 1\frac{8}{15} =$ $\qquad$ $7\frac{5}{12} + 3\frac{2}{9} + 4\frac{3}{4} =$

7. In 1990 the town of Midvale spent $\$1\frac{1}{4}$ million on education. In 1995 it spent $\$2\frac{1}{2}$ million more on education than it spent in 1990. How much did it spend on education in 1995?

8. The porch behind Clark's house was $5\frac{1}{2}$ feet wide. Clark built a $6\frac{7}{12}$-foot wide extension on the porch. How wide was the new porch?

9. One weekend Jeff put paneling on the walls of his dining room. Friday he worked $2\frac{1}{4}$ hours. Saturday he worked $6\frac{2}{3}$ hours. Sunday he worked $3\frac{1}{2}$ hours. What was the total number of hours he worked that weekend?

10. Celeste's empty suitcase weighs $6\frac{1}{4}$ pounds. She packed the suitcase with $15\frac{9}{16}$ pounds of clothes. How much did the suitcase weigh when it was full of clothes?

11. It usually takes John $\frac{3}{4}$ of an hour to drive to work. One day because of flooding on the highway, it took him $1\frac{1}{2}$ hours longer than usual. How long did it take John to drive to work that day?

12. Andy is $71\frac{1}{2}$ inches tall. His brother Fred is $1\frac{3}{4}$ inches taller. How many inches tall is Fred?

To check your answers, turn to page 183.

Subtraction of Fractions

Pat works $7\frac{1}{2}$ hours a day. She eats lunch after she has been at work for $3\frac{1}{4}$ hours. How many hours does she have to work after lunch?

To subtract fractions, subtract the numerators and put the difference (the answer) over the denominator. When fractions in do not have the same denominators, first find a common denominator. Change each fraction to a new fraction with the common denominator. Then subtract.

Example: $7\frac{1}{2} - 3\frac{1}{4} =$

STEP 1. Find the LCD. 2 and 4 both divide evenly into 4. 4 is the LCD.

STEP 2. Raise $\frac{1}{2}$ to a new fraction with 4 as the denominator.

STEP 3. Subtract the new fractions.

STEP 4. Subtract the whole numbers.

$$\begin{array}{r} 7\frac{1}{2} = 7\frac{2}{4} \\ -\,3\frac{1}{4} = 3\frac{1}{4} \\ \hline 4\frac{1}{4} \end{array}$$

➠ Pat has to work $4\frac{1}{4}$ hours after lunch.

PRACTICE 31

Rewrite each problem, subtract, and reduce.

1. $\frac{7}{8} - \frac{1}{8} =$ $\qquad \frac{3}{5} - \frac{3}{10} =$ $\qquad 8\frac{2}{3} - 1\frac{1}{4} =$ $\qquad 7\frac{5}{9} - 3\frac{1}{3} =$

2. $\frac{4}{5} - \frac{1}{5} =$ $\qquad \frac{6}{7} - \frac{1}{3} =$ $\qquad 9\frac{3}{4} - 6\frac{3}{8} =$ $\qquad 6\frac{4}{5} - 4\frac{2}{3} =$

3. Abdul bought a piece of lumber $85\frac{1}{2}$ inches long. He sawed off a piece that was $32\frac{1}{8}$ inches long. How long was the piece that was left?

4. Jennie has a piece of plastic tubing that is $18\frac{5}{16}$ inches long. She needs a piece that measures $10\frac{3}{4}$ inches. How much will she have left if she cuts the length she needs?

To check your answers, turn to page 183.

Subtraction with Regrouping

To repair a broken chair, Leon cut a piece of wood $6\frac{7}{8}$ inches long from a piece that was 12 inches long. How long was the remaining piece?

Sometimes there is no top fraction to subtract the bottom fraction from. To get something in the position of the top fraction, you must **regroup** or borrow. To regroup means to write the whole number on top as a whole number and an improper fraction. For example, $12 = 11\frac{8}{8}$. The numerator and denominator of the improper fraction should be the same as the denominator of the other fraction in the problem.

Example: $12 - 6\frac{7}{8} =$

STEP 1. Regroup 12 as $11\frac{8}{8}$ since 8 is the common denominator.

STEP 2. Subtract the fractions. $\frac{8}{8} - \frac{7}{8} = \frac{1}{8}$.

STEP 3. Subtract the whole numbers. $11 - 6 = 5$.

$$\begin{array}{r} 12 = 11\frac{8}{8} \\ -6\frac{7}{8} = 6\frac{7}{8} \\ \hline 5\frac{1}{8} \end{array}$$

Remember that any fraction with the same numerator and denominator is equal to 1.

➡ The remaining piece is $5\frac{1}{8}$ inches long.

Sometimes the top fraction is not big enough to subtract the bottom fraction from. To get a bigger fraction in the top position, regroup the mixed number at the top. Study the next example carefully.

Example: $5\frac{1}{8} - 2\frac{7}{8} =$

STEP 1. Regroup 5 as $4\frac{8}{8}$.

STEP 2. Add $\frac{8}{8}$ to $\frac{1}{8}$. $\frac{8}{8} + \frac{1}{8} = \frac{9}{8}$.

STEP 3. Subtract the new fractions.

STEP 4. Subtract the new whole numbers.

STEP 5. Reduce the answer.

$$\begin{array}{r} 5\frac{1}{8} = 4\frac{8}{8} + \frac{1}{8} = 4\frac{9}{8} \\ -2\frac{7}{8} = \phantom{4\frac{8}{8} + \frac{1}{8} =} 2\frac{7}{8} \\ \hline 2\frac{2}{8} = 2\frac{1}{4} \end{array}$$

PRACTICE 32

A. Subtract and reduce each problem.

1. $6 - 2\frac{3}{7} =$ $\qquad 9 - 7\frac{7}{8} =$ $\qquad 5 - 1\frac{7}{9} =$ $\qquad 4 - 1\frac{3}{4} =$

2. $9\frac{3}{7} - 4\frac{6}{7} =$ $\qquad 5\frac{4}{9} - 1\frac{7}{9} =$ $\qquad 8\frac{2}{5} - 3\frac{4}{5} =$ $\qquad 7\frac{1}{4} - 2\frac{3}{4} =$

When the denominators in a subtraction problem are not the same, change each fraction to a new fraction with a common denominator. Then regroup if you need to.

Example: $8\frac{1}{3} - 4\frac{7}{9} =$

STEP 1. Find the LCD. 3 divides evenly into 9. 9 is the LCD.

STEP 2. Raise $\frac{1}{3}$ to a fraction with 9 as the denominator. $\qquad \frac{1}{3} = \frac{3}{9}$

STEP 3. Regroup 8 as $7\frac{9}{9}$.

STEP 4. Add $\frac{9}{9}$ and $\frac{3}{9}$.

STEP 5. Subtract the new fractions.

STEP 6. Subtract the whole numbers.

STEP 7. Reduce the answer. $3\frac{5}{9}$ is reduced.

$$\begin{array}{r} 8\frac{1}{3} = 8\frac{3}{9} = 7\frac{9}{9} + \frac{3}{9} = 7\frac{12}{9} \\ -\ 4\frac{7}{9} = 4\frac{7}{9} = \qquad\qquad 4\frac{7}{9} \\ \hline 3\frac{5}{9} \end{array}$$

B. Subtract and reduce.

3. $8\frac{1}{2} - 5\frac{2}{3} =$ $\qquad 7\frac{2}{5} - 2\frac{5}{6} =$ $\qquad 3\frac{1}{4} - 1\frac{5}{6} =$ $\qquad 7\frac{1}{3} - 3\frac{5}{8}$

4. $12\frac{1}{2} - 4\frac{7}{9} =$ $\qquad 6\frac{1}{2} - 4\frac{7}{8} =$ $\qquad 5\frac{2}{9} - 1\frac{5}{6} =$ $\qquad 9\frac{5}{12} - 4\frac{3}{4} =$

5. The Midvale City Council needs \$4 million to build a new sports arena. So far the council has raised $\$1\frac{7}{8}$ million. How much more does the council need to raise?

To check your answers, turn to page 183.

Multiplication of Fractions

At the last election in Northport, $\frac{1}{2}$ of the eligible voters went to the polls. Of the voters who showed up, $\frac{3}{5}$ of them voted to re-elect the mayor. What fraction of the eligible voters voted for the current mayor?

When you multiply whole numbers (except 1 and 0), the answer is *bigger* than the two numbers you multiply. When you multiply two proper fractions, the answer is *smaller* than either of the two fractions. When you multiply two fractions, you find **a part of a part.**

For example, if you multiply $\frac{1}{2}$ by $\frac{1}{2}$, you find $\frac{1}{2}$ *of* $\frac{1}{2}$. You know that $\frac{1}{2}$ of $\frac{1}{2}$ dollar is $\frac{1}{4}$ dollar. The answer is smaller than either of the fractions you multiplied.

$\frac{1}{2}$ of [50¢ coin] = [25¢ coin]

50¢ 25¢

$\frac{1}{2}$ of $\frac{1}{2}$ dollar = $\frac{1}{4}$ dollar

To multiply fractions, multiply the numerators together and the denominators. Then reduce the answer.

Example: $\frac{3}{5} \times \frac{1}{2} =$

STEP 1. Multiply the numerators: $3 \times 1 = 3$.

STEP 2. Multiply the denominators: $5 \times 2 = 10$.

STEP 3. Try to reduce. $\frac{3}{10}$ is reduced.

$$\frac{3}{5} \times \frac{1}{2} = \frac{3}{10}$$

➡ $\frac{3}{10}$ of the eligible voters voted for the current mayor.

PRACTICE 33

Multiply and reduce.

1. $\frac{1}{5} \times \frac{2}{3} =$ $\frac{3}{8} \times \frac{3}{5} =$ $\frac{4}{5} \times \frac{3}{7} =$ $\frac{1}{3} \times \frac{4}{5} =$

2. $\frac{5}{6} \times \frac{1}{8} =$ $\frac{3}{4} \times \frac{7}{10} =$ $\frac{4}{9} \times \frac{2}{5} =$ $\frac{1}{7} \times \frac{1}{8} =$

To check your answers, turn to page 184.

Canceling

Canceling is a way of making multiplication of fractions problems easier. Canceling is similar to reducing. To cancel, divide a numerator and a denominator by a number that goes evenly into both of them.

Example: Multiply $\frac{8}{9} \times \frac{3}{20} =$

Step 1. Cancel: Divide 4 into 8 and 20. Cross out 8 and write 2. Cross out 20 and write 5.

$$\frac{\overset{2}{\cancel{8}}}{9} \times \frac{3}{\underset{5}{\cancel{20}}} =$$

Step 2. Cancel: Divide 3 into 3 and 9. Cross out 3 and write 1. Cross out 9 and write 3.

$$\frac{\overset{2}{\cancel{8}}}{\underset{3}{\cancel{9}}} \times \frac{\overset{1}{\cancel{3}}}{\underset{5}{\cancel{20}}}$$

Step 3. Multiply the new numerators.

Step 4. Multiply the new denominators.

$$\frac{\overset{2}{\cancel{8}}}{\underset{3}{\cancel{9}}} \times \frac{\overset{1}{\cancel{3}}}{\underset{5}{\cancel{20}}} = \frac{2}{15}$$

Step 5. Try to reduce. $\frac{2}{15}$ is reduced.

PRACTICE 34

Cancel, multiply, and reduce.

1. $\frac{4}{9} \times \frac{6}{7} =$ $\qquad \frac{5}{6} \times \frac{3}{7} =$ $\qquad \frac{2}{3} \times \frac{7}{8} =$ $\qquad \frac{3}{4} \times \frac{8}{15}$

2. $\frac{5}{9} \times \frac{3}{20} =$ $\qquad \frac{4}{9} \times \frac{3}{8} =$ $\qquad \frac{7}{10} \times \frac{8}{21} =$ $\qquad \frac{5}{12} \times \frac{6}{13} =$

3. $\frac{5}{18} \times \frac{4}{15} =$ $\qquad \frac{8}{9} \times \frac{9}{10} =$ $\qquad \frac{8}{15} \times \frac{9}{14} =$ $\qquad \frac{4}{21} \times \frac{9}{16} =$

4. $\frac{4}{9} \times \frac{5}{14} =$ $\qquad \frac{8}{33} \times \frac{11}{24} =$ $\qquad \frac{9}{10} \times \frac{4}{9} =$ $\qquad \frac{6}{7} \times \frac{3}{8} =$

To check your answers, turn to page 184.

Multiplication with Fractions and Whole Numbers

Roberto works at a car repair shop. He estimates that he spends $\frac{5}{6}$ of his time changing oil filters. On a day when he works 9 hours, how much time does he spend changing oil filters.

To multiply a whole number and a fraction, first write the whole number as a fraction. Write the whole number as the numerator and 1 as the denominator.

Example: Multiply $\frac{5}{6} \times 9 =$

STEP 1. Write 9 as a fraction with a denominator of 1. $\frac{5}{6} \times \frac{9}{1}$

STEP 2. Cancel: Divide 9 and 6 by 3. $\frac{5}{\cancel{6}_{2}} \times \frac{\cancel{9}^{3}}{1} = \frac{15}{2} = 7\frac{1}{2}$

STEP 3. Multiply the new numerators.

STEP 4. Multiply the new denominators.

STEP 5. Change the answer, $\frac{15}{2}$, to a mixed number, $7\frac{1}{2}$.

To review changing improper fractions to mixed numbers, turn to page 57.

➡ Roberto spends $7\frac{1}{2}$ hours changing oil filters.

PRACTICE 35

Cancel, multiply, and reduce. Change every improper fraction answer to a mixed number.

1. $8 \times \frac{3}{4} =$ $\quad 15 \times \frac{2}{5} =$ $\quad \frac{5}{9} \times 12 =$ $\quad \frac{1}{6} \times 9 =$

2. $\frac{4}{5} \times 10 =$ $\quad \frac{5}{12} \times 20 =$ $\quad 6 \times \frac{2}{3} =$ $\quad 5 \times \frac{3}{10} =$

3. $2 \times \frac{7}{8} =$ $\quad 7 \times \frac{5}{14} =$ $\quad \frac{3}{8} \times 16 =$ $\quad \frac{8}{15} \times 18 =$

To check your answers, turn to page 184.

Multiplication with Mixed Numbers

Bill has $4\frac{1}{2}$ gallons of paint left from painting his house. He will need $\frac{1}{3}$ of this amount to paint his garage. How much paint will the garage require?

To multiply mixed numbers, first change the mixed numbers to improper fractions. Then multiply the improper fractions.

Example: $\frac{1}{3} \times 4\frac{1}{2} =$

STEP 1. Change $4\frac{1}{2}$ to an improper fraction: $4\frac{1}{2} = \frac{9}{2}$.

$$\frac{1}{3} \times 4\frac{1}{2} = \frac{1}{3} \times \frac{9}{2}$$

To review changing mixed numbers to improper fractions, turn to page 58.

STEP 2. Cancel: Divide 3 and 9 by 3.

STEP 3. Multiply the new numerators: $1 \times 3 = 3$.

$$\frac{1}{\cancel{3}_1} \times \frac{\cancel{9}^3}{2} = \frac{3}{2} = 1\frac{1}{2}$$

STEP 4. Multiply the new denominators: $1 \times 2 = 2$.

STEP 5. Change $\frac{3}{2}$ to a mixed number. $\frac{3}{2} = 1\frac{1}{2}$.

To review changing improper fractions to mixed numbers, turn to page 57.

➠ The garage requires $1\frac{1}{2}$ gallons.

PRACTICE 36

Multiply each problem. Change improper fraction answers to mixed numbers.

1. $\frac{8}{9} \times 3\frac{3}{4} =$ $\quad 2\frac{1}{12} \times \frac{9}{10} =$ $\quad 6\frac{2}{3} \times \frac{15}{16} =$ $\quad \frac{4}{9} \times 2\frac{5}{8} =$

2. $4\frac{2}{3} \times 6 =$ $\quad 12 \times 1\frac{3}{4} =$ $\quad 1\frac{3}{10} \times 5 =$ $\quad 9 \times 2\frac{1}{6} =$

3. $\frac{2}{3} \times 4\frac{4}{5} =$ $\quad 2\frac{4}{7} \times \frac{7}{9} =$ $\quad \frac{3}{4} \times 2\frac{2}{9} =$ $\quad 3\frac{1}{5} \times \frac{5}{6} =$

4. $4\frac{2}{3} \times \frac{6}{7} =$ $\quad \frac{3}{8} \times 3\frac{1}{5} =$ $\quad 6\frac{2}{3} \times \frac{3}{4} =$ $\quad \frac{9}{20} \times 5\frac{1}{3}$

5. $\frac{5}{6} \times 1\frac{5}{7} =$ $1\frac{7}{8} \times \frac{4}{9} =$ $\frac{5}{7} \times 1\frac{3}{10} =$ $2\frac{2}{5} \times \frac{7}{8} =$

6. One can of tomato sauce weighs $4\frac{3}{4}$ ounces. What is the total weight of the cans in a box that holds 16 cans of tomato sauce?

7. Gordy bought $5\frac{1}{4}$ yards of material at $6.80 a yard. How much did he pay for the material? [**Hint:** Remember to put two decimal places in problems involving dollars and cents.]

To check your answers, turn to page 184.

Division by Fractions

A pie recipe calls for $\frac{1}{4}$ cup of flour. Colleen has $\frac{1}{2}$ cup of flour. The flour she has is enough for how many pies?

To find the number of pies Colleen can make, divide $\frac{1}{2}$ by $\frac{1}{4}$. To divide fractions, **invert** the divisor (the number at the right of the ÷ sign) and follow the rules for multiplying fractions. To invert means to turn a fraction around. Write the numerator on the bottom and the denominator on the top. For example, when you invert $\frac{2}{3}$, you get $\frac{3}{2}$.

In fraction division problems, the amount being divided comes first.

Example: $\frac{1}{2} \div \frac{1}{4} =$

STEP 1. Invert the divisor $\frac{1}{4}$ to $\frac{4}{1}$ and change the division sign (÷) to a multiplication sign (×)

$$\frac{1}{2} \div \frac{1}{4} = \frac{1}{2} \times \frac{4}{1}$$

STEP 2. Cancel: Divide 2 and 4 by 2.

STEP 3. Multiply the new numerators: 1 × 2 = 2.

$$\frac{1}{\cancel{2}_{1}} \times \frac{\cancel{4}^{2}}{1} = \frac{2}{1} = 2$$

STEP 4. Multiply the new denominators: 1 × 1 = 1.

STEP 5. Change the improper fraction to a whole number: $\frac{2}{1} = 2$.

Colleen has enough flour for two pies.

PRACTICE 37

A. Divide each problem. Change every improper fraction answer to a whole or mixed number.

1. $\frac{9}{10} \div \frac{3}{4} =$ $\frac{1}{2} \div \frac{3}{8} =$ $\frac{6}{7} \div \frac{3}{5} =$ $\frac{5}{8} \div \frac{1}{12} =$

2. $\frac{4}{9} \div \frac{2}{3} =$ $\frac{1}{4} \div \frac{5}{12} =$ $\frac{5}{8} \div \frac{15}{16} =$ $\frac{3}{10} \div \frac{2}{5} =$

In fraction division problems, change whole numbers and mixed numbers to improper fractions.

B. Divide each problem.

3. $6 \div \frac{3}{8} =$ $8 \div \frac{4}{5} =$ $9 \div \frac{6}{7} =$ $5 \div \frac{2}{3} =$

4. $4 \div \frac{1}{2} =$ $3 \div \frac{9}{10} =$ $2 \div \frac{1}{4} =$ $7 \div \frac{7}{10} =$

5. $3\frac{1}{2} \div \frac{3}{8} =$ $4\frac{4}{5} \div \frac{2}{5} =$ $7\frac{1}{2} \div \frac{9}{16} =$ $1\frac{3}{4} \div \frac{1}{2} =$

6. $1\frac{1}{3} \div \frac{5}{9} =$ $2\frac{2}{9} \div \frac{5}{12} =$ $3\frac{1}{2} \div \frac{7}{10} =$ $5\frac{1}{4} \div \frac{7}{8} =$

7. Sam is filling plastic containers with wood screws. Each container holds $\frac{1}{16}$ pound of screws. How many containers can he fill with $\frac{7}{8}$ pound of wood screws?

8. How many stakes, each $\frac{2}{3}$ foot long, can be cut from a pole that is twelve feet long?

9. Andrea needs $\frac{3}{4}$ yard of material to make an apron. How many aprons can she make from $3\frac{1}{2}$ yards of material?

To check your answers, turn to page 185.

Division of Fractions and Mixed Numbers by Whole Numbers

Mary and Mack want to share $\frac{7}{8}$ pound of blueberries equally. Find the weight of each of their shares.

When you divide a fraction or a mixed number by a whole number, you "split" the fraction or mixed number into smaller parts. For example, if you divide a fraction by 2, you split the fraction into two equal parts. To divide a fraction by a whole number, first write the whole number as a fraction. Write the whole number as the numerator and 1 as the denominator. Then invert the fraction you are dividing by and multiply. Follow the rules for multiplying fractions.

Example: Divide $\frac{7}{8} \div 2 =$

Step 1.	Write 2 as a fraction with a denominator of 1.	$\frac{7}{8} \div 2 = \frac{7}{8} \div \frac{2}{1}$
Step 2.	Invert the divisor $\frac{2}{1}$ to $\frac{1}{2}$ and change the division sign ($\div$) to a multiplication sign ($\times$).	$\frac{7}{8} \div \frac{2}{1} = \frac{7}{8} \times \frac{1}{2}$
Step 3.	Multiply the numerators. $7 \times 1 = 7$	$\frac{7}{8} \times \frac{1}{2} = \frac{7}{16}$
Step 4.	Multiply the denominators. $8 \times 2 = 16$	
Step 5.	Try to reduce. $\frac{7}{16}$ is reduced.	

➡ Mary and Mack each get $\frac{7}{16}$ pound of blueberries.

PRACTICE 38

Divide each problem.

1. $\frac{3}{5} \div 6 =$ $\quad$ $\frac{4}{9} \div 10 =$ $\quad$ $\frac{7}{8} \div 14 =$ $\quad$ $\frac{1}{4} \div 3 =$

2. $1\frac{1}{5} \div 4 =$ $\quad$ $2\frac{1}{2} \div 5 =$ $\quad$ $1\frac{2}{3} \div 10 =$ $\quad$ $2\frac{2}{5} \div 2 =$

3. One month the Randolphs made 9 long distance calls. The total time for the calls was $56\frac{1}{4}$ minutes. What was the average time for each call? [**Hint:** Divide the total time by the number of calls.]

4. George has $3\frac{1}{2}$ ounces of a chemical. He needs to divide the chemical equally into 5 test tubes. How much should he put in each test tube?

To check your answers, turn to page 185.

Division by Mixed Numbers

A carpenter needs $4\frac{2}{3}$ yards of wood to make a cabinet. How many cabinets can he make from 30 yards of wood?

To divide by a mixed number, first change the mixed number to an improper fraction. Then invert the fraction you are dividing by and follow the rules for multiplying fractions.

Example: $30 \div 4\frac{2}{3} =$

STEP 1. Change both 30 and $4\frac{2}{3}$ to improper fractions.

$$30 \div 4\frac{2}{3} = \frac{30}{1} \div \frac{14}{3}$$

STEP 2. Invert the divisor $\frac{14}{3}$ to $\frac{3}{14}$ and change the division sign to a multiplication sign.

$$\frac{30}{1} \div \frac{14}{3} = \frac{30}{1} \times \frac{3}{14}$$

STEP 3. Cancel: Divide 30 and 14 by 2.

STEP 4. Multiply the new numerators. $15 \times 3 = 45$

$$\frac{\overset{15}{\cancel{30}}}{1} \times \frac{3}{\underset{7}{\cancel{14}}} = \frac{45}{7} = 6\frac{3}{7}$$

STEP 5. Multiply the new denominators: $1 \times 7 = 7$

STEP 6. Change the answer, $\frac{45}{7}$, to a mixed number.

The answer, $6\frac{3}{7}$ cabinets, does not make sense. There is no such thing as a fraction of a cabinet.

➡ The carpenter can make 6 cabinets from the wood.

PRACTICE 39

Divide each problem.

1. $4 \div 1\frac{3}{5} =$ $\quad 3 \div 4\frac{1}{2} =$ $\quad 9 \div 3\frac{3}{4} =$ $\quad 6 \div 1\frac{2}{7} =$

2. $14 \div 3\frac{1}{2} =$ $\quad 5 \div 1\frac{3}{7} =$ $\quad 6 \div 2\frac{2}{3} =$ $\quad 7 \div 4\frac{1}{5} =$

3. $8 \div 1\frac{1}{3} =$ $\quad 2 \div 1\frac{3}{4} =$ $\quad 3 \div 1\frac{5}{7} =$ $\quad 20 \div 3\frac{1}{5} =$

In these problems change both mixed numbers to improper fractions. Then invert the fraction you are dividing by.

4. $3\frac{3}{4} \div 2\frac{1}{2} =$ $\quad 1\frac{7}{8} \div 2\frac{1}{4} =$ $\quad 5\frac{1}{2} \div 2\frac{1}{16} =$ $\quad 2\frac{5}{8} \div 1\frac{3}{4} =$

5. $6\frac{1}{2} \div 9\frac{3}{4} =$ $\quad 4\frac{4}{5} \div 1\frac{1}{15} =$ $\quad 8\frac{1}{3} \div 3\frac{1}{3} =$ $\quad 3\frac{5}{9} \div 4\frac{4}{9} =$

6. Harriet paid \$4.05 for $2\frac{1}{4}$ pounds of ground beef. Find the cost of one pound of beef.

7. Selma wants to cut a piece of copper tubing $67\frac{1}{2}$ inches long into $7\frac{1}{2}$-inch pieces. How many pieces can she cut from the tube?

8. Angelo paid \$16.50 for $3\frac{2}{3}$ yards of material. Find the price of one yard of the material.

9. Jim is building a brick walk from his driveway to his front door. Each brick is $3\frac{1}{2}$ inches wide. He wants the walk to be 49 inches wide. How many rows of bricks does he need?

To check your answers, turn to page 185.

Fractions Review

These problems will help you find out if you need to review the fractions section of this book. Solve each problem. When you finish, look at the chart to see which pages you should review.

1. There are 1,000 meters in a kilometer. 13 meters are what fraction of a kilometer?

2. Circle the proper fractions: $\frac{9}{9}$ $3\frac{1}{2}$ $\frac{4}{11}$ $\frac{9}{2}$ $\frac{3}{8}$ $\frac{6}{5}$

3. Circle the improper fractions: $\frac{9}{2}$ $\frac{4}{7}$ $8\frac{1}{2}$ $\frac{2}{11}$ $\frac{6}{6}$ $\frac{13}{3}$

4. Circle the mixed numbers: $\frac{3}{3}$ $4\frac{2}{15}$ $\frac{15}{2}$ $\frac{1}{6}$ $8\frac{4}{7}$ $\frac{2}{9}$

5. Reduce $\frac{25}{45}$ to lowest terms.

6. Raise $\frac{4}{7}$ to an equal fraction with a denominator of 56.

7. Change $\frac{40}{6}$ to a mixed number and reduce.

8. Change $4\frac{5}{6}$ to an improper fraction.

9. Which fraction is greater, $\frac{5}{8}$ or $\frac{7}{10}$?

10. $\frac{3}{7} + \frac{2}{7} =$

11. $\frac{8}{15} + \frac{4}{5} =$

12. $3\frac{2}{3} + 4\frac{5}{8} + 5\frac{1}{4} =$

13. $\frac{3}{5} - \frac{1}{6} =$

14. $9 - 4\frac{5}{12} =$

15. $3\frac{5}{6} - \frac{1}{3} =$

16. Sam sawed a piece of lumber $15\frac{5}{8}$ inches long from a board 36 inches long. How long was the piece leftover?

17. $\frac{5}{12} \times \frac{4}{5} =$

18. $8 \times \frac{7}{10} =$

19. $1\frac{7}{8} \times 3\frac{1}{5} =$

20. The Jordan family makes $27,900 a year. They spend $\frac{1}{3}$ of their income for food and clothing. How much do they spend on food and clothing in a year?

21. $\frac{3}{10} \div \frac{2}{5} =$

22. $\frac{7}{8} \div 3 =$

23. $1\frac{1}{3} \div 2\frac{2}{3} =$

24. Paul paid $15.40 for $4\frac{2}{3}$ yards of pipe. How much did one foot of pipe cost?

PROGRESS CHECK

Check your answers on page 186. Then return to the review pages for the problems you missed. Correct your answers before going on to the next unit.

If you missed problems	*Review pages*
1 to 4	63 to 57
5 to 9	58 to 62
10 to 12	63 to 65
13 to 16	66 to 68
17 to 20	69 to 73
21 to 24	74 to 78

UNIT 4

Ratio and Proportion

Writing and Simplifying Ratios

Sue makes \$16 an hour at Premier Plumbing. When she started working at Premier, she made \$9 an hour. What is the ratio of her wage now to her starting wage?

A **ratio** is a comparison of numbers by division. A ratio can be written with the word *to,* with a colon (:), or as a fraction. The fraction method is most common. The first number in a ratio is the numerator (top number). The second number is the denominator (bottom number).

Example: Write a ratio of Sue's current wage of \$16 an hour to her starting wage of \$9 an hour.

STEP 1. Write the first number mentioned, 16, as the numerator. $\frac{16}{\ }$

STEP 2. Write the second number, 9, as the denominator. $\frac{16}{9}$

➡ The ratio of Sue's current wage to her starting wage is $\frac{16}{9}$ or 16:9 or 16 to 9.

PRACTICE 40

A. Solve each problem.

1. Write the ratio of 2 pounds of hamburger for 3 people as a fraction.

2. Write the ratio of 5 runs in 9 innings with a colon (:).

3. Write the ratio of 16 students for 1 teacher with the word *to*.

Like a fraction, a ratio can be simplified or reduced. To simplify a ratio, find a number that divides evenly into both the numerator and the denominator.

Remember that simplifying changes the numbers in a ratio, but it does not change the value.

Example: Simplify the ratio 9 to 6.

STEP 1. Divide both 9 and 6 by a number that divides evenly into both. 3 divides evenly into 9 and 6.

$$\frac{9 \div 3}{6 \div 3} = \frac{3}{2}$$

To review reducing, turn to page 54.

STEP 2. Check to see if another number, besides, 1, divides evenly into 3 and 2. No other number divides evenly into both. The ratio 3:2 is simplified as far as it can be.

B. For problems 4–6, simplify each ratio.

4. 20:25

5. 27 to 36

6. $\frac{28}{35}$

For problems 7–8 write each ratio with a colon (:) and simplify.

7. $24 for 2 tickets

8. 400 pieces in 12 hours

Use the following information for problems 9–11.

So far Hanna has paid $1,500 on her car loan. She still has to pay $1,000.

9. What is the ratio of the amount Hanna has paid to the amount she still has to pay?

10. What is the ratio of the amount Hanna still has to pay to the amount she has already paid?

12. Write a ratio for the amount Hanna has paid on her loan to the total amount of the loan.

To check your answers, turn to page 186.

Ratio of Measurement

Sandy works as a receptionist. She spends 20 minutes out of every hour talking on the phone. What is the ratio of the time she spends on the phone to the total time she spends at work?

To compare units of measurement, express them in the same terms. To compare minutes and hours, change both to minutes.

Example: Simplify the ratio of 20 minutes to one hour.

STEP 1.	Change 1 hour to minutes. One hour = 60 minutes.	20 min:1 hr = 20:60
STEP 2.	Divide 20 and 60 by a number that divides evenly into both, 20.	20:60 = 1:3

➡ The ratio of the time Sandy spends talking on the phone to the total time she spends at work is 1:3.

PRACTICE 41

Use the table of time measurements to solve problems 1–5.

Time
1 minutes (min) = 60 seconds (sec)
1 hour (hr) = 60 min
1 day (da) = 24 hr
1 week (wk) = 7 days
1 year (yr) = 365 days = 12 months

Write each ratio with a colon (:) and simplify.

1. 24 minutes to an hour

2. 9 months to 1 year

3. 4 hours to a day

4. 6 days to 2 weeks

5. During an 8-hour work day, Kyle estimates that he spends 6 hours waiting on customers. What is the ratio of time spent helping customers to his total work day?

Use the table of length measurements to solve problems 6–11.

Length
1 foot (ft) = 12 inches (in.)
1 yard (yd) = 3 ft = 36 in.
1 mile (mi) = 5,280 ft = 1,760 yd

Write each ratio as a fraction and simplify.

6. 4 inches to 1 foot

7. 2 feet to 20 inches

8. 2,640 feet to 1 mile

9. 4 yards to 6 feet

10. Jane's workbench is 30 inches wide and eight feet long. What is the ratio of the width to the length?

11. A room is 26 feet long and 5 yards wide. What is the ratio of the length to the width?

Use the table of liquid measurements to solve problems 12–16.

Liquid Measure
1 cup = 8 fluid ounces (fl oz)
1 pint (pt) = 2 cups = 16 fl oz
1 quart (qt) = 2 pints = 32 fl oz
1 gallon (gal) = 4 quarts

Write each ratio with the word *to* and simplify.

12. 3 quarts to 1 gallon

13. 10 fluid ounces to 1 pint

14. 1 pint to 6 cups

15. 1 cup to 4 fluid ounces

16. A full bathtub uses about 80 quarts of water. A 10-minute shower uses about 40 gallons of water. Write a ratio comparing the water used to take a bath to the water used to take a 10-minute shower.

Use the table of weight measurements to solve problems 17–19.

Weight
1 pound (lb) = 16 ounces (oz)
1 ton = 2,000 pounds

Write each ratio with a colon (:) and simplify.

17. 10 ounces to 1 pound

18. 1 ton to 400 pounds

19. Sam's truck weighs 2 tons. He hauled kitchen appliances that weighed a total of 800 pounds. What is the ratio of the weight of the appliances to the weight of his truck?

To check your answers, turn to page 186.

Writing Proportions

A **proportion** is a statement that two ratios are equal. The statement $\frac{6}{8} = \frac{3}{4}$ is a proportion. A proportion is the same as two equal fractions. Each of the four numbers in a proportion (or a pair of equal fractions) is called a **term.**

The **cross products** in a proportion are equal. This means that the top term of the first ratio times the bottom term of the second equals the bottom term of the first ratio times the top term of the second. For the proportion $\frac{6}{8} = \frac{3}{4}$, both cross products equal 24.

$$\frac{6}{8} = \frac{3}{4}$$

$$6 \times 4 = 8 \times 3$$

$$24 = 24$$

PRACTICE 42

For problems 1–2, write the terms in each cross product. Then write the answer to each cross product. The first proportion is written as an example.

1. $\frac{3}{5} = \frac{9}{15}$ $\frac{8}{3} = \frac{16}{6}$ $\frac{2}{5} = \frac{10}{25}$

   $3 \times 15 = 5 \times 9$

   $45 = 45$

2. $\frac{12}{9} = \frac{16}{12}$ $\frac{1}{4} = \frac{20}{80}$ $\frac{15}{10} = \frac{3}{2}$

For problems 3–4, first rewrite each proportion with fractions. Then write the terms and the answer to each cross product.

3. 6:24 = 2:8   20:4 = 30:6   3:2 = 33:22

   $\frac{6}{24} = \frac{2}{8}$

   $6 \times 8 = 24 \times 2$

   $48 = 48$

4. 8:28 = 4:14   3:10 = 9:30   15:9 = 10:6

To check your answers, turn to page 187.

Solving Proportions

The ratio of office workers to shop workers at Sarma's job is 2:7. There are 14 office workers at Sarma's job. How many shop workers are there?

Proportion is a useful way to solve many word problems. In most proportion problems, one of the four terms is missing. Use a letter to stand for the missing term.

When a term is missing in a proportion, only one cross product can be completed. The other cross product is incomplete. To find the missing term, divide the completed cross product by the number in the incomplete cross product.

Example: Find the number of shop workers at Sarma's job.

STEP 1. Set up the proportion carefully. Write labels to identify the terms. Notice that the number of office workers, 14, goes on top. Use the letter *s* to represent shop workers.

$$\frac{\text{office}}{\text{shop}} \quad \frac{2}{7} = \frac{14}{s}$$

STEP 2. Cross multiply. The cross product of 2 and *s* is $2 \times s$. The cross product of 7 and 14 is 98.

$$2 \times s = 7 \times 14$$
$$2 \times s = 98$$

STEP 3. To find *s*, divide 98 by the number in the incomplete cross product, 2.

$$s = 98 \div 2$$
$$s = 49$$

➡ There are 49 shop workers at Sarma's job.

PRACTICE 43

A. Solve each problem.

In problems 1–2, solve each proportion for the letter *n*.

1. $\frac{12}{9} = \frac{8}{n}$ $\qquad$ $\frac{2}{7} = \frac{n}{28}$ $\qquad$ $\frac{16}{n} = \frac{2}{3}$

2. $\frac{n}{15} = \frac{4}{6}$ $\qquad$ $\frac{3}{5} = \frac{9}{n}$ $\qquad$ $\frac{14}{3} = \frac{n}{6}$

For problem 3, rewrite each proportion as two fractions. Then solve for the letter *x*.

3. 8:11 = 2:c 5:2 = c:7 6:c = 5:3

For word problems, remember that the parts in a proportion must correspond to each other.

Example: A four-foot pipe weighs fifteen pounds. What is the weight of a seven-foot pipe?

STEP 1. Write a proportion with the ratio of feet (length) to pounds (weight). Use w for the weight of the 7-foot pipe.

$$\frac{\text{feet}}{\text{pounds}} \quad \frac{4}{15} = \frac{7}{w}$$

STEP 2. Cross multiply. The cross product of 4 and w is $4 \times w$. The cross product of 15 and 7 is 105.

$$4 \times w = 15 \times 7$$
$$4 \times w = 105$$

STEP 3. Divide 105 by the number in the incomplete cross product, 4.

$$w = 105 \div 4$$
$$w = 26\frac{1}{4}$$

➡ The seven-foot pipe weighs $26\frac{1}{4}$ pounds.

B. Solve each problem.

4. The ratio of games won to games lost for the Westport Warriors was 5:2. The team won 15 games. How many games did they lose?

5. A six-acre field produced 180 bushels of wheat. At the same rate of production, how many bushels of wheat can be produced on 10 acres?

6. Blanca took a photograph that was 3 inches wide and 5 inches long to be enlarged so that it was 20 inches long. If the width and length are in the same proportion as the original, how wide is the enlargement?

7. A cake recipe calls for 4 cups of sugar and 3 tablespoons of butter. If Jack reduced the amount of sugar to 3 cups, how much butter should he use?

To check your answers, turn to page 187.

Proportion Shortcuts

The shop where Carmela works produces backpacks. On an average, 7 out of every 200 backpacks are defective. If the shop produces 1,200 backpacks a year, how many of them are defective?

Like many proportion problems, this one has rather large numbers. Instead of finding cross products and dividing, write out each cross product and *cancel* by the number in the incomplete cross product.

Example: Find the number of defective backpacks produced in a year at Carmela's shop.

STEP 1. Set up a proportion with ratios of defective backpacks to the total number of backpacks produced. $\frac{\text{defective}}{\text{total}} \quad \frac{7}{200} = \frac{d}{1{,}200}$

STEP 2. Write the cross products. $200 \times d = 7 \times 1{,}200$

STEP 3. To solve for *d,* write 200 below the cross product 7 × 1,200. Then cancel. $d = \frac{7 \times \overset{6}{\cancel{1{,}200}}}{\underset{1}{\cancel{200}}}$

➡ Out of 1,200 backpacks, 42 are defective. $d = 42$

Notice that the fraction bar below 7 × 1,200 acts as a division sign. You are dividing 200 into 7 × 1,200.

PRACTICE 44

Solve each problem.

1. $\frac{30}{50} = \frac{m}{35}$ $\qquad \frac{9}{400} = \frac{36}{m}$ $\qquad \frac{800}{m} = \frac{400}{15}$

2. $m{:}420 = 4{:}21$ $\qquad 8{:}25 = m{:}200$ $\qquad 60{:}100 = 48{:}m$

3. A train traveled 178 miles in two hours. At the same rate, how far can the train travel in 5 hours?

4. Shaquil paid $38.40 for three gallons of boat varnish. Find the cost of ten gallons of boat varnish.

5. On a road map, two inches equal 15 miles. Find the distance in miles between two towns that are six inches apart on the map.

6. At Premier Plumbing, 7 out of 10 workers got raises at their last review. Altogether there are 80 employees at Premier. How many of them got raises?

7. At Midvale Elementary School, 17 out of every 70 students ride a bus to school. If there are 700 students at the school, how many come to school by bus?

8. Of every $100 John earns, he pays $22 in taxes. During a 5-month period, John will earn $10,000. How much of his income during that period will go to taxes?

Use the following information to answer questions 9–11.

Chantal mixed one gallon of white paint with three gallons of yellow paint to get the color she wanted for her house.

9. What is the total number of gallons in the paint mixture?

10. What is the ratio of white paint to the total number of gallons in the mixture?

11. Chantal used a total of 24 gallons of paint on her house. How many gallons of white paint did she use?

To check your answers, turn to page 188.

Ratio and Proportion Review

These problems will help you decide if you need to review the ratio and proportion unit of this book. When you finish, look at the chart to see which pages you should review.

For problems 1–3, simplify each ratio.

1. 30:36

2. 24 to 16

3. $\frac{45}{27}$

Use the following information for problems 4–5.

The Jays won 66 games and lost 22.

4. What is the ratio of the games won to the games lost?

5. What is the ratio of the games won to the games played?

6. Write and simplify the ratio of 8 months to 1 year.

7. Write and simplify the ratio of 40 inches to 1 yard.

For problems 8–10, solve each proportion for the missing number (n).

8. $\frac{5}{9} = \frac{15}{n}$

9. $\frac{30}{7} = \frac{n}{2}$

10. $n:8 = 400:50$

11. Arlene drove 115 miles in two hours. Driving at the same rate, how far can she go in three hours?

12. The ratio of defective cans to the total number of cans produced at Container, Inc. is 3:500. The company produces 30,000 cans a month. How many of them are defective?

PROGRESS CHECK

Check your answers on page 189. Then return to the review pages for the problems you missed. Correct your answers before going on to the next unit.

If you missed problems	*Review pages*
1 to 7	81 to 84
8 to 12	85 to 89

UNIT 5

Percent

Writing Percent

Like a fraction or a decimal, a percent shows a part of a whole. Fractions divide a whole into 2 parts or 3 parts or 4 parts and so on. Decimals divide a whole into 10 parts or 100 parts, or 1,000 parts and so on. The word *percent* means "out of 100." Percents divide a whole into 100 parts and only 100 parts. A percent is very much like a two-place decimal (hundredths).

The figure pictured here is divided into 100 parts, and 70 of the parts are shaded. The fraction $\frac{70}{100}$ which reduces to $\frac{7}{10}$ tells how much of the figure is shaded. The decimal .70 which simplifies to .7 also tells how much of the figure is shaded. 70% of the figure is shaded.

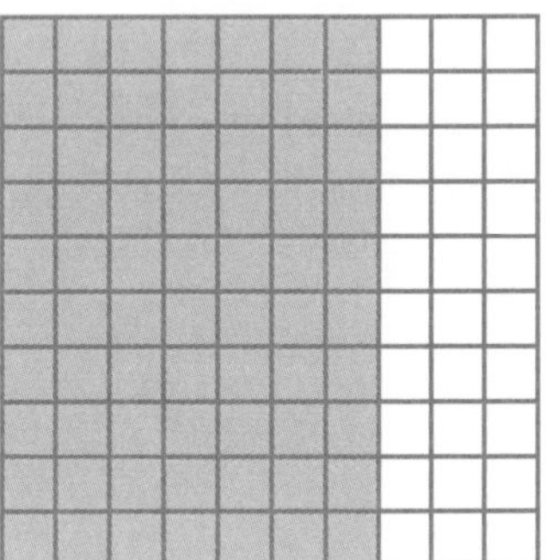

To work with percent, think of a whole as 100%

Example: Jordan spends 35% of his income for rent. What percent of his income is left for everything else?

100% represents Jordan's total income. To find the percent that goes for everything besides rent, subtract 35% from 100%.

$$\begin{array}{r} 100\% \\ -\ 35\% \\ \hline 65\% \end{array}$$

➡ Jordan has 65% of his income left after paying rent.

PRACTICE 45

A. Answer each question.

1. The Rodriguez family has paid off 75% of their mortgage. What percent of the mortgage do they still have to pay?

2. One evening during a flu epidemic, 30% of Marla's students were absent from her exercise class. What percent of the students came to class?

3. Junior got 82% of the problems right on his last math test. What percent of the problems did he get wrong?

4. 12% of the employees at the Municipal Hospital walk to work. What percent arrive by some means other than walking?

Remember that one whole is 100%. If something increases, it becomes more than 100% of itself.

Example: In the fifteen years the Johnsons have owned their house, the value of the house has doubled. The value now is what percent of the value fifteen years ago?

Remember that one whole is 100%. If something doubles, it becomes 2 × 100%.

$2 \times 100\% = 200\%$

➡ The Johnsons' house is worth 200% of its value fifteen years ago.

B. Answer each question.

5. Serena makes four times what she made when she first got out of school. Her income now is what percent of her income when she first left school?

6. The current population of Eastport is three times what it was in 1960. The population now is what percent of the 1960 population?

To check your answers, turn to page 190.

Percents and Decimals

To work with percent, you should be able to convert decimals, fractions, and percents to corresponding forms. The next three lessons explain how to interchange these forms.

It is easy to change decimals to percents. Move the decimal point two places to the **right** and write a percent sign.

Example: Change .35 to a percent.

STEP 1. Move the point two places to the right. $.35 = .35 = 35\%$

STEP 2. Write a percent sign.

Notice that when the decimal point moves to the end, you do not have to write it.

Example: Change .052 to a percent.

STEP 1. Move the point two places to the right. $.052 = .052 = 5.2\%$

STEP 2. Write a percent sign.

Example: Change $.66\frac{2}{3}$ to a percent.

STEP 1. Move the point two places to the right. $.66\frac{2}{3} = .66\frac{2}{3} = 66\frac{2}{3}\%$

STEP 2. Write a percent sign.

Notice that when the decimal point comes just before a fraction, you do not have to write it.

Sometimes you will have to put zeros after the decimal to get two places.

Example: Change .7 to a percent.

STEP 1. Put a zero at the right of .7 $.7 = .70 = 70\%$

STEP 2. Move the point two places to the right.

STEP 3. Write a percent sign.

PRACTICE 46

A. Change each decimal to a percent.

1. $.65 =$ $\qquad$ $.06 =$ $\qquad$ $.045 =$ $\qquad$ $.16\frac{2}{3} =$

2. $.8 =$ $\qquad$ $.25 =$ $\qquad$ $.06\frac{1}{4} =$ $\qquad$ $.82 =$

3. $2.8 =$ $\qquad$ $.5 =$ $\qquad$ $.625 =$ $\qquad$ $4. =$

To change a percent to a decimal, move the decimal point two places to the **left.** Then take off the percent sign.

Example: Change 65% to a decimal.

STEP 1. Move the point two places to the left. 65% = 65% = .65

STEP 2. Take off the percent sign.

Sometimes you will have to put zeros at the left to get two places.

Example: Change 2.8% to a decimal.

STEP 1. Put a zero to the left of 2.8%. 2.8% = 02.8% = .028

STEP 2. Move the point two places to the left.

STEP 3. Take off the percent sign.

Example: Change 20% to a decimal.

STEP 1. Move the point two places to the left.

STEP 2. Take off the percent sign. 20% = 20% = .2

Notice that we took off the zero at the right in .20. The zero does not change the value of .2.

Example: Change $44\frac{4}{9}\%$ to a decimal.

STEP 1. The point is understood to be at the right of 44. Move the point two places to the left.

$$44\frac{4}{9}\% = 44\frac{4}{9}\% = .44\frac{4}{9}$$

STEP 2. Take off the percent sign.

B. Change each percent to a decimal.

4. 55% =	8% =	12.5% =	2% =
5. 6.4% =	$33\frac{1}{3}\%$ =	60% =	225% =
6. 90% =	0.4% =	20% =	500% =

To check your answers, turn to page 190.

Percents and Fractions

Percents are different from common fractions in two ways. One difference is that 100 is the only number that can be a denominator for percents. The other difference is that the denominator is not written. Instead of writing 100, we write the % sign. There are two different ways to change a fraction to a percent. Look at the examples on this page carefully. Then choose the way you like better.

One method for changing a fraction to a percent is to multiply the fraction by 100%.

Example: Change $\frac{2}{5}$ to a percent.

Multiply $\frac{2}{5}$ by 100%

$$\frac{2}{5} \times 100\% = \frac{2}{\cancel{5}_{1}} \times \frac{\overset{20\%}{\cancel{100\%}}}{1} = \frac{40\%}{1} = 40\%$$

The other method is to change the fraction to a decimal first. Then change the decimal to a percent.

Example: Change $\frac{3}{4}$ to a percent.

Step 1. Change $\frac{3}{4}$ to a decimal. $\frac{3}{4} = 4\overline{)3.00}$ = .75

Step 2. Change .75 to a percent. .75 = .75 = 75%

PRACTICE 47

A. Change each fraction to a percent.

1.	$\frac{7}{10} =$	$\frac{3}{5} =$	$\frac{1}{9} =$	$\frac{9}{50} =$
2.	$\frac{8}{25} =$	$\frac{3}{8} =$	$\frac{1}{3} =$	$\frac{1}{2} =$
3.	$\frac{5}{12} =$	$\frac{9}{20} =$	$\frac{1}{16} =$	$\frac{1}{4} =$

Remember that a percent is a kind of fraction. To change a percent to a fraction, write the digits in the percent as the numerator. Write 100 as the denominator. Then reduce the fraction.

Example: Change 85% to a fraction.

Step 1. Write 85 as the numerator and 100 as the denominator. $\frac{85}{100} = \frac{17}{20}$

Step 2. Reduce the fraction by 5.

When a percent has a decimal in it, first change the percent to a decimal. Then change the decimal to a fraction and reduce.

Example: Change 8.4% to a fraction.

Step 1. Change 8.4% to a decimal. 8.4% = 08.4 = .084

Step 2. Change .084 to a fraction. $\frac{84}{1,000} = \frac{21}{250}$

Step 3. Reduce the fraction by 4.

When a percent has a fraction in it, write the digits in the percent as the numerator. Write 100 as the denominator. Then **divide** the numerator by the denominator. This operation is hard. Study the following example carefully.

Example: Change $58\frac{1}{3}\%$ to a fraction.

STEP 1. Write $58\frac{1}{3}$ as the numerator and 100 as the denominator.

$$\frac{58\frac{1}{3}}{100}$$

STEP 2. Remember that the line separating the numerator from the denominator means **divided by.** Divide $58\frac{1}{3}$ by 100.

$$58\frac{1}{3} \div 100 =$$

$$\frac{175}{3} \div \frac{100}{1} =$$

$$\frac{\overset{7}{\cancel{175}}}{3} \times \frac{1}{\underset{4}{\cancel{100}}} = \frac{7}{12}$$

B. Change each percent to a fraction and reduce.

4. 35% = 2% = 24% = 30% =

5. 44% = 6% = 150% = 3% =

6. 4.8% = 10.5% = .04% = 2.75% =

7. $12\frac{1}{2}\%$ = $83\frac{1}{3}\%$ = $42\frac{6}{7}\%$ = $8\frac{1}{3}\%$ =

8. 90% 215% 6.4% $22\frac{2}{9}\%$

To check your answers, turn to page 190.

Common Fractions, Decimals, and Percents

The chart on this page includes some of the fractions, decimals, and percents you will use most often in your work. You will learn that sometimes it is easier to change a percent to a decimal. Sometimes it is easier to change a percent to a fraction. Fill in the chart. Then check your answers. Make sure your answers are correct. Then study the list carefully. Memorize the equivalent fraction and decimal for each percent.

PRACTICE 48

Fill in the blanks.

Percent	Decimal	Fraction
50%	______	______
25%	______	______
75%	______	______
$12\frac{1}{2}\%$	______	______
$37\frac{1}{2}\%$	______	______
$62\frac{1}{2}\%$	______	______
$87\frac{1}{2}\%$	______	______
$33\frac{1}{3}\%$	______	______
$66\frac{2}{3}\%$	______	______

Percent	Decimal	Fraction
20%	______	______
40%	______	______
60%	______	______
80%	______	______
10%	______	______
30%	______	______
70%	______	______
90%	______	______
$16\frac{2}{3}\%$	______	______
$83\frac{1}{3}\%$	______	______

To check your answers, turn to page 191.

Identifying the Percent, the Whole, and the Part

Anne calculated her rent increase. 5% of her current rent of $480 a month is $24.

Anne's situation is a typical percent problem. It has a percent, a whole, and a part.

The **percent** is easy to identify. It has the % sign. The word *of* suggests multiplication. The word that follows *of* is usually the **whole.** (For Anne the whole is her monthly rent.) The verb *is* suggests the = sign. The product of multiplying the percent times the whole is the **part.**

Example: Identify the percent, the whole, and the part in the statement: 5% of $480 is $24.

%	whole	part
5%	$480	$24

PRACTICE 49

Fill in the blanks with the percent, the whole, and the part for each statement.

	%	whole	part
1. 50% of 88 is 44.	______	______	______
2. 12 is 25% of 48.	______	______	______
3. 90% of $300 is $270.	______	______	______
4. On a test with 20 problems, Maxim got 75% right. He got 15 problems right.	______	______	______

To check your answers, turn to page 191.

Finding the Part

At the Peerless Package Company, 20% of the 35 employees work part-time. How many of the employees work part-time?

In the last lesson you learned that *percent* × *whole* = *part*. For Peerless Package, the percent is 20% and the whole is the 35 employees.

To multiply by a percent, first change the percent to a decimal or a fraction. In this lesson you can practice both methods.

Example: Use a decimal to find 20% of 35.

To review changing percents to decimals, turn to page 92.

STEP 1. Change 20% to a decimal. 20% = 20% = .2

STEP 2. Multiply to find the part. Multiply 35 by .2.

$$\begin{array}{r} 35 \\ \times\ .2 \\ \hline 7.0 \end{array} = 7$$

➡ There are 7 part-time employees at Peerless.

PRACTICE 50

A. Use decimals to solve each problem.

1. 75% of 48 = 30% of 60 = 10% of 18 =

2. 35% of 80 = 16% of 300 = 21% of 400 =

3. 12.5% of 56 = 6.5% of 400 = 1.2% of 800 =

4. Mr. and Mrs. Caruso make $32,500 a year. They put 12% of their income into a retirement plan. How much do they put in their retirement plan in a year?

5. The sales tax in Jermaine's state is 6%. She bought a coat for $85. How much was the sales tax on the coat?

You can also change the percent to a fraction to solve for the part.

Example: Use a fraction to find 20% of 40.

STEP 1. Change 20% to a fraction. $20\% = \frac{10}{100} = \frac{1}{5}$

To review changing percents to fractions, turn to page 94.

STEP 2. Multiply 35 by $\frac{1}{5}$. $\frac{1}{\cancel{5}_{1}} \times \frac{\cancel{35}^{7}}{1} = \frac{7}{1} = 7$

➡ 20% of 40 is 7.

B. Use fractions to solve each problem. Use any fraction equivalents from the table on page 97 that you need.

6. 60% of 80 = 25% of 32 = 30% of 50 =

7. 20% of 55 = 75% of 48 = 50% of 92 =

8. 35% of 180 = $66\frac{2}{3}$% of 150 = $12\frac{1}{2}$% of 40 =

9. Ching Mae's English test had 30 questions. She got 90% of them right. How many questions did she get right?

10. Raul weighed 144 pounds. He started working out with weights and gained $12\frac{1}{2}$% of his weight. How many pounds did he gain?

11. Gloria makes $630 a week. Her employer takes out 20% of her pay for taxes and social security. How much does Gloria's employer take out each week?

To check your answers, turn to page 192.

Using Proportion to Find the Part

The plant where Alexis works makes hammers. 18% of the hammers are shipped outside the country. One month the factory produced 400 hammers. How many were shipped outside the country?

Besides first changing a percent to a decimal or changing a percent to a fraction, there is another way to solve percent problems. Set up a proportion using the model shown here. Then solve the proportion.

There is no single right way to solve percent problems. This lesson gives you a chance to practice using proportion.

$$\frac{\text{part}}{\text{whole}} = \frac{\%}{100}$$

With a proportion, you do not need to change the percent to a decimal or to a fraction.

Example: Use a proportion to find 18% of 400.

STEP 1. Set up a proportion. 18 is the percent, and 400 is the whole. You are looking for the part. Use n for the part.

$$\frac{n}{400} = \frac{18}{100}$$

STEP 2. Find the cross products and solve.

$$n = \frac{\overset{4}{\cancel{400}} \times 18}{\underset{1}{\cancel{100}}}$$

$$n = 72$$

To review proportion shortcuts, turn to page 86.

➠ 72 hammers were shipped outside the country.

PRACTICE 51

Use proportion to solve each problem.

1. 6% of 250 = 75% of 84 = 30% of 130 =

2. 37.5% of 72 = 4.5% of 500 = 12% of 600 =

3. 60% of 380 = 50% of 204 = 15% of 7,000 =

4. When Kate bought her house, it was worth $90,000. Now it is worth 140% of the price Kate paid. What is the value of the house now?

5. 1,500 people were interviewed in a recent poll. 65% of them approved of the President's foreign policies. How many of them approved?

6. The Jays have played 15 games and won 60% of them. How many games have they won?

7. There are 45 employees in Arlene's office. 40% of them have worked in the office less than one year. How many of the employees have worked there less than a year?

To check your answers, turn to page 192.

Multi-Step Problems

In some percent problems, you have to use two operations to find an answer. First find a percent of a number. Then add or subtract this new amount with the original amount in the problem.

Example: Dave bought a coat on sale. The coat used to cost $65. He bought it for 20% off the old price. How much did Dave pay for the coat?

STEP 1. Change 20% to a fraction.

$$20\% = \frac{20}{100} = \frac{1}{5}$$

STEP 2. Multiply $65 by $\frac{1}{5}$.

$$\frac{1}{\cancel{5}_{1}} \times \frac{\cancel{65}^{13}}{1} = \$13$$

STEP 3. Subtract $13 from $65.

$$\$65 - \$13 = \$52$$

➡ Dave paid $52 for the coat.

PRACTICE 52

Read each problem carefully to decide if you need to add or subtract.

1. Selma makes \$560 a week. Her employer takes out 15% of her pay for taxes and Social Security. How much does Selma take home each week?

2. Frank took a math test with 40 problems. He got 85% of the problems right. How many problems did he get wrong?

3. Don bought tapes for \$15.60. The sales tax in his state is 5%. How much did the tapes cost including sales tax?

4. In June Petra's phone bill was \$24.50. In July her bill was 30% more because of long distance calls. How much was her July phone bill?

5. On a normal work day about 24,000 people ride the buses in Midvale. Monday was a holiday, and the number of riders was down 35%. How many people rode the buses on Monday?

To check your answers, turn to page 192.

Interest

Interest is money someone pays for using someone else's money. A bank pays you interest for using your money in a savings account. You pay a bank interest for using the bank's money on a loan.

To find interest, multiply the principal by the rate by the time.

The **principal** is the money you borrow or save.
The **rate** is the percent of the interest.
The **time** is the number of years.

Example: Find the interest on $900 at 7% annual interest for one year.

STEP 1. Change 7% to a fraction. $7\% = \frac{7}{100}$

STEP 2. Multiply the principal by the rate by the time. $\frac{\overset{9}{\cancel{900}}}{1} \times \frac{7}{\underset{1}{\cancel{100}}} \times 1 = \$63$

PRACTICE 53

A. Find the interest for each of the following.

1. $800 at 6% annual interest for one year. $600 at 3.5% annual interest for one year.

2. $400 at 12% annual interest for one year. $450 at 4.8% annual interest for one year.

When the time period for interest is not one year, change the time to a fraction of a year.

Example: Find the interest on $900 at 7% annual interest for 8 months.

STEP 1. Change 7% to a fraction. $7\% = \frac{7}{100}$

STEP 2. Change 8 months to a fraction. Write 8 in the numerator and 12 months (one whole year) in the denominator. $\frac{8}{12} = \frac{2}{3}$

STEP 3. Multiply the principal by the rate by the time. $\frac{\overset{3}{\cancel{\overset{9}{\cancel{900}}}}}{1} \times \frac{7}{\underset{1}{\cancel{100}}} \times \frac{2}{\underset{1}{\cancel{3}}} = \$42$

When the time period is more than one year, write the time as a mixed number. For example, one year and six months = $1\frac{6}{12} = 1\frac{1}{2}$.

B. Find the interest for each of the following.

3. $480 at 5% annual interest for 5 months. $2,400 at 10% annual interest for 1 year and 8 months.

4. $360 at 3.5% annual interest for 1 year and 4 months.

$3,600 at 9% annual interest for 1 year and 9 months.

To check your answers, turn to page 193.

Finding the Percent

Max works 40 hours a week as a landscaper. He spends eight hours every week maintaining equipment. What percent of his time at work is spent maintaining equipment?

When you studied fractions, you learned how to find what part one number is of another. You made a fraction with the part as the numerator (top number) and the whole as the denominator (bottom number). The steps are the same for finding what percent one number is of another. Make a fraction with the part over the whole. Then change the fraction to a percent.

Example: 8 is what percent of 40?

STEP 1. Make a fraction with the part, 8, over the whole, 40, and reduce.

$$\frac{\text{part}}{\text{whole}} \quad \frac{8}{40} = \frac{1}{5}$$

STEP 2. Change $\frac{1}{5}$ to a percent.

$$\frac{1}{\cancel{5}_{1}} \times \frac{\cancel{100\%}^{20}}{1} = \frac{20}{1} = 20\%$$

➠ Max spends 20% of his work time maintaining equipment.

PRACTICE 54

Solve each problem.

1. 12 is what percent of 48?

   What percent of 32 is 16?

2. What percent of 45 is 36?

   14 is what percent of 21?

3. 16 is what percent of 160?

   18 is what percent of 45?

4. What percent of 50 is 15? 30 is what percent of 48?

5. The Mejias make $2,200 a month. They spend $660 a month for food. What percent of their income do the Mejias spend on food?

6. Last year Marvin weighed 220 pounds. He went on a diet and lost 22 pounds. What percent of his weight did he lose?

7. 350 people work at the Allied Paper Products factory. 210 workers at the factory are part of the volunteer savings plan. What percent of the workers participate in the savings plan?

To check your answers, turn to page 193.

Using Proportion to Find the Percent

Donald spends ten hours of every work day away from home. He spends two of those hours driving. What percent of his time away from home does Donald spend driving?

Proportion is a convenient way to set up percent problems. Use the model shown here.

$$\frac{\text{part}}{\text{whole}} = \frac{\%}{100}$$

Example: Use proportion to solve the problem, 2 is what percent of 10?

STEP 1. Set up a proportion. 2 is the part, and 10 is the whole. You are looking for the percent. Use p for the percent.

$$\frac{2}{10} = \frac{p}{100}$$

Step 2. Find the cross products and solve.

$$p = \frac{2 \times \cancel{100}^{10}}{\cancel{10}_{1}}$$

$$p = 20\%$$

To review proportion shortcuts, turn to page 86.

➡ Donald spends 20% of his time away from home driving.

PRACTICE 55

Use proportion to solve each problem.

1. 15 is what percent of 75? What percent of 48 is 30?

2. 60 is what percent of 90? 70 is what percent of 200?

3. What percent of 230 is 23? 95 is what percent of 190?

4. 126 is what percent of 140? 400 is what percent of 320?

5. 27 is what percent of 300? What percent of 200 is 125?

6. Wallis took a test with 50 problems. She got 42 problems right. What percent of the problems did Wallis get right?

7. Jonelle borrowed $4,000. She had to pay $440 interest on the amount she borrowed. The interest was what percent of the loan?

8. Norman works 20 hours a week for a food co-op. He spends about 12 of those hours making deliveries. What percent of his time working for the co-op is spent making deliveries?

To check your answers, turn to page 193.

Multi-Step Problems

In some problems you have to compare the difference between two amounts to an original amount. First subtract to find the difference. Then make a fraction with the difference as the numerator and the original amount as the denominator. Change the fraction to a percent.

Example: Last year Jane paid \$480 a month for rent. This year she pays \$552 a month. By what percent did Jane's rent increase?

STEP 1. Find how much Jane's rent went up. Subtract the old rent from the new rent.

$$\begin{array}{r} \$552 \\ -\ 480 \\ \hline \$\ 72 \end{array}$$

STEP 2. Make a fraction with the difference, \$72, over the original rent, \$480, and reduce.

$$\frac{72}{480} = \frac{3}{20}$$

STEP 3. Change $\frac{3}{20}$ to a percent.

$$\frac{3}{\cancel{20}_{1}} \times \frac{\cancel{100}^{9}}{1} = 15\%$$

➠ Jane's rent went up 15%.

PRACTICE 56

Solve each problem.

1. Last year Nancy made \$16.20 an hour. This year she makes \$17.82 an hour. By what percent did her wage increase?

2. Mr. Walek runs a shoe store. He pays an average of \$45 for a pair of shoes. He charges his customers an average of \$63. By what percent does he mark up the price of a pair of shoes?

3. Two years ago Larry bought gas for \$1.20 gallon. This year he pays \$1.26 a gallon. By what percent did the price go up?

4. Jeff bought a T.V. on sale for $187. Before the sale the T.V. cost $220. Find the percent of discount on the original price.

5. Last year there were 16 students in Mr. Green's night school math class. This year there are 22 students. By what percent did the number of students increase?

To check your answers, turn to page 194.

Finding the Whole

There are 12 mechanics working at Ted's Tire Company. Mechanics are 80% of the total number of employees. How many employees does Ted's have?

Finding the whole is "backwards" from finding the part. For Ted's Tire Company, you are looking for the number of employees that, when multiplied by 80%, gives 12. First change the percent to a fraction or a decimal. Then **divide** the part by the fraction or decimal.

Example: 80% of what number is 12?

Using a fraction:

STEP 1. Change 80% to a fraction. $80\% = \frac{80}{100} = \frac{4}{5}$

STEP 2. Divide 12 by $\frac{4}{5}$. $12 \div \frac{4}{5} =$

$$\frac{\overset{3}{\cancel{12}}}{1} \times \frac{5}{\underset{1}{\cancel{4}}} = \frac{15}{1} = 15$$

Using a decimal:

STEP 1. Change 80% to a decimal. $80\% = 80\% = .8$

STEP 2. Divide 12 by .8.

$$8\overline{)120}\quad\text{quotient } 15$$

➡ There are 15 employees at Ted's Tire Co.

It is a good idea to check these "backwards" problems. To check the example, find 40% of 40. The answer should be 16.

Using a fraction: $80\% = \frac{4}{5}$

$$\frac{4}{\cancel{5}_1} \times \frac{\cancel{15}^3}{1} = \frac{12}{1} = 12$$

Using a decimal: $80\% = .8$

$$.8 \times 15 = 12.0 = 12$$

PRACTICE 57

Use fractions or decimals to solve these problems.

1. 60% of what number is 27? $12\frac{1}{2}\%$ of what number is 8?

2. 50% of what number is 14? 60% of what number is 24?

3. 25% of what number is 40? $16\frac{2}{3}\%$ of what number is 25?

4. 40% of what number is 52? $37\frac{1}{2}\%$ of what number is 45?

5. $83\frac{1}{3}\%$ of what number is 35? 75% of what number is 150?

6. 70% of the members of a carpenters' union voted to strike. 210 members voted to strike. How many members are there in the union?

7. Marcus pays $190 a month on his car loan. This 10% of his monthly income. Find his monthly income.

8. Maya got 48 questions right on a Spanish test. Her score was 80%. How many questions were on the test?

To check your answers, turn to page 194.

Using Proportion to Find the Whole

In a recent vote only 30% of the employees at Ace Electronics said they would join a union. Of all the workers, 84 voted to join a union. How many employees are there at Ace?

Again, proportion is a useful way to set up percent problems. Use the model shown here. Remember that you are looking for the whole, the total number of employees at Ace Electronics.

$$\frac{\text{part}}{\text{whole}} = \frac{\%}{100}$$

Example: Use proportion to solve the problem, 30% of what number is 84?

STEP 1. Set up a proportion. 84 is the part, and 30 is the percent. You are looking for the whole. Use w for the whole.

$$\frac{84}{w} = \frac{30}{100}$$

To review proportion short-cuts, turn to page 86.

STEP 2. Find the cross products and solve.

$$w = \frac{\overset{28}{\cancel{84}} \times \overset{10}{\cancel{100}}}{\underset{1}{\cancel{\underset{\cancel{3}}{30}}}}$$

$$w = 280$$

➡ There are 280 employees at Ace Electronics.

PRACTICE 58

Use proportion to solve each problem.

1. 75% of what number is 15? 40% of what number is 24?

2. 20% of what number is 35? 15% of what number is 36?

3. 8% of what number is 28? 95% of what number is 380?

4. The Hawks lost 28 games last season. They lost 35% of the games they played. How many games did the Hawks play?

To check your answers, turn to page 194.

PERCENT REVIEW

These problems will help you find out if you need to review the percent section of this book. Solve each problem. When you finish, look at the chart to see which pages you should review.

1. Change .09 to a percent.

2. Change 48% to a decimal.

3. Change $\frac{5}{12}$ to a percent.

4. Change 85% to a fraction.

5. Find 15% of 125.

6. What is 370% of 90?

7. What is 4.8% of 800?

8. Find $66\frac{2}{3}\%$ of 129.

9. The Lopez family makes $28,600 a year. They spend 25% of their income on mortgage payments. How much do they spend in a year on mortgage payments?

10. Evan bought a shirt for $29.80. He had to pay 5% sales tax. How much did Evan pay for the shirt including tax?

11. Find the interest on $1,600 at 5% annual interest for one year.

12. Find the interest on $720 at 15% annual interest for nine months.

13. 45 is what percent of 75

14. 36 is what percent of 54?

15. 48 is what percent of 192?

16. 60 is what percent of 96?

17. Last year Fred made $460 a week. This year he got a raise of $23 a week. His raise is what percent of his old weekly salary?

18. Before Joe went on a diet, he weighed 180 pounds. Now he weighs 153 pounds. What percent of his weight did he lose?

19. 30% of what number is 57?

20. 75% of what number is 108?

21. 48% of what number is 60?

22. $33\frac{1}{3}$% of what number is 24?

23. There were 240,000 people living in Central County in 1980. That was 75% of the number living there in 1995. How many people lived in Central County in 1995?

PROGRESS CHECK

Check your answers on page 194. Then return to the review pages for the problems you missed. Correct your answers before going on to the next unit.

If you missed problems	*Review pages*
1 to 4	90 to 97
5 to 10	99 to 102
11 to 12	103 to 104
13 to 18	105 to 108
19 to 23	109 to 111

UNIT

6

Algebra

Writing Algebra

Algebra is another tool to use in solving mathematical problems. Algebra is arithmetic written with letters as well as numbers.

There are four basic number operations: addition, subtraction, multiplication, and division. The answer to an addition problem is called the **sum.** The plus sign (+) shows addition. The answer to a subtraction problem is called the **difference.** The minus sign (−) means to subtract.

The answer to a multiplication problem is the **product.** In arithmetic, the times sign (×) means to multiply. In algebra, multiplication is shown using a raised dot (·) or by writing two numbers or symbols next to each other.

arithmetic	**algebra**
3×4	$3 \cdot 4$ or $3(4)$

The answer to a division problem is the **quotient.** In arithmetic, the signs ÷ or $\overline{)}$ mean to divide. In algebra, division is shown with the fraction bar.

arithmetic	**algebra**
$8 \div 2$ or $2\overline{)8}$	$\frac{8}{2}$

Perhaps the most important symbol in arithmetic or algebra is the equal sign (=). In mathematical statements, it often stands for the word *is*.

PRACTICE 59

Write each of the following with numbers and symbols. The first one is done as an example.

1. the sum of eight and seven $8 + 7$
2. the product of nine and five ______
3. the sum of one-half and three ______
4. the quotient of ten divided by two ______
5. The sum of twelve and three is fifteen. ______

To check your answers, turn to page 195.

Powers

Besides addition, subtraction, multiplication, and division, algebra uses an operation called *raising a number to a power.*

8^2 means "eight to the second power." The number 8 is called the **base.** The small number 2 is called the **exponent.**

$$\text{base} \rightarrow 8^2 \leftarrow \text{exponent}$$

To find a power:

1. Write the base the number of times the exponent tells you.
2. Multiply the base by itself.

Raising a number to the second power is also called *squaring* a number. 8^2 is *eight to the second power* or *eight squared.*

Example: Find the value of 8^2.

STEP 1.	The exponent is 2. Write 8 two times.	$8^2 = 8 \times 8$
STEP 2.	Multiply 8 by 8.	$8 \times 8 = 64$

➡ $8^2 = 64$.

Raising a number to the third power is also called *cubing* a number. 5^3 is *five to the third power* or *five cubed.*

Example: Find the value of 5^3.

STEP 1.	The exponent is 3. Write 5 three times.	$5^3 = 5 \times 5 \times 5$
STEP 2.	Multiply 5 by 5. Then multiply that product, 25, by 5.	$5 \times 5 \times 5 = 125$

➡ $5^3 = 125$

PRACTICE 60

A. Solve each problem.

1. $10^2 =$ $\qquad$ $6^2 =$ $\qquad$ $12^2 =$

2. $4^3 =$ $\qquad$ $15^2 =$ $\qquad$ $2^4 =$

3. $(.3)^2 =$ $\qquad$ $\left(\frac{1}{2}\right)^2 =$ $\qquad$ $(.01)^2 =$

To use powers with addition and subtraction, first calculate each power separately. Then add or subtract the values from left to right.

Example: Find the value of $3^2 + 4^2$.

STEP 1.	Find the value of each power.	$3^2 = 3 \times 3 = 9$ $4^2 = 4 \times 4 = 16$
STEP 2.	Add the values of each power.	$9 + 16 = 25$

B. Find the value of each each of the following.

4. $6^2 + 3^2 =$ $\qquad$ $8^2 - 2^2 =$ $\qquad$ $10^3 - 5^2 =$

5. $8^2 - 5^2 =$ $\qquad$ $4^2 - 3^2 + 5^2 =$ $\qquad$ $10^2 - 9^2 =$

To check your answers, turn to page 195.

Square Roots

The opposite of powers is called **roots.** The opposite of finding a second power is finding a **square root.** For example, 5 to the second power is 25. The square root of 25 is 5. The symbol for square root is $\sqrt{\ }$. The statement $\sqrt{25} = 5$ means "the square root of 25 is 5."

To find a square root, find a number that multiplied by itself gives the number inside the $\sqrt{\ }$ sign.

Example: Find $\sqrt{49}$.

Ask yourself, "What number times itself is 49?" You know that $7 \times 7 = 49$.

$\sqrt{49} = 7$

➡ The square root of 49 is 7.

When you square whole numbers, the answers are *perfect squares.* For example, $3^2 = 9$, $7^2 = 49$, and so on. If you know the most common perfect squares, it is easy to find the square roots.

Square Roots of Perfect Squares

$\sqrt{1} = 1$	$\sqrt{36} = 6$	$\sqrt{121} = 11$
$\sqrt{4} = 2$	$\sqrt{49} = 7$	$\sqrt{144} = 12$
$\sqrt{9} = 3$	$\sqrt{64} = 8$	$\sqrt{169} = 13$
$\sqrt{16} = 4$	$\sqrt{81} = 9$	$\sqrt{196} = 14$
$\sqrt{25} = 5$	$\sqrt{100} = 10$	$\sqrt{225} = 15$

PRACTICE 61

Find each square root or power. Then add or subtract the values.

1. $\sqrt{64} + \sqrt{9} =$ $\quad\quad$ $\sqrt{36} - \sqrt{4} =$ $\quad\quad$ $\sqrt{100} - \sqrt{25} =$

2. $\sqrt{16} + 5^2 =$ $\quad\quad$ $9^2 - \sqrt{9} =$ $\quad\quad$ $\sqrt{144} - 1^2 =$

To check your answers, turn to page 195.

The Number Line

All the whole numbers, decimals, and fractions that are *greater than zero* are **positive numbers.** The number line shown here represents the positive numbers.

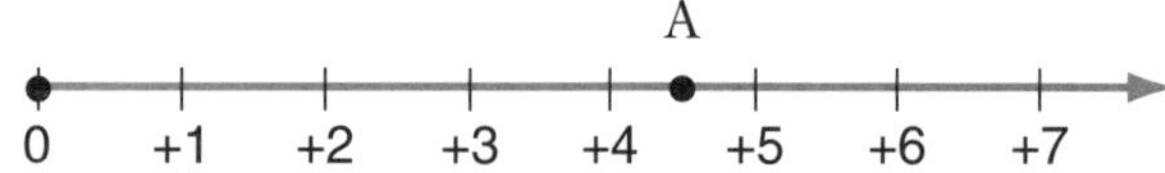

The dot at zero is the starting point. Every point on the line to the right of zero has value greater than zero—the farther to the right, the greater the value. The arrow means that the numbers continue.

The point labeled *A* is half-way between 4 and 5. Point *A* represents 4.5 or $4\frac{1}{2}$.

Algebra sometimes uses numbers that are *less than zero*. These are called **negative numbers.** The number line shown below represents both positive numbers and negative numbers.

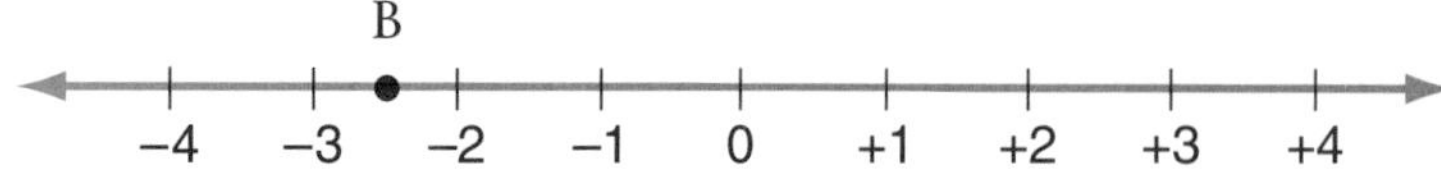

Points to the right of zero have a value greater than zero. Points to the left of zero have a value less than zero. The arrows mean that the numbers continue in both directions.

Zero has no sign. It is neither positive nor negative.

The point labeled *B* is half-way between -2 and -3. Point *B* represents -2.5 or $-2\frac{1}{2}$. This number number is read as "minus two and one-half" or "negative two and one-half."

The + sign means to add. It also means a positive (or plus) number. If a number has no sign, it is *understood* to be positive. The − sign means to subtract. It also means a negative (or minus) number.

PRACTICE 62

Use the number line below to identify the value of each labeled point. The first one is done as an example.

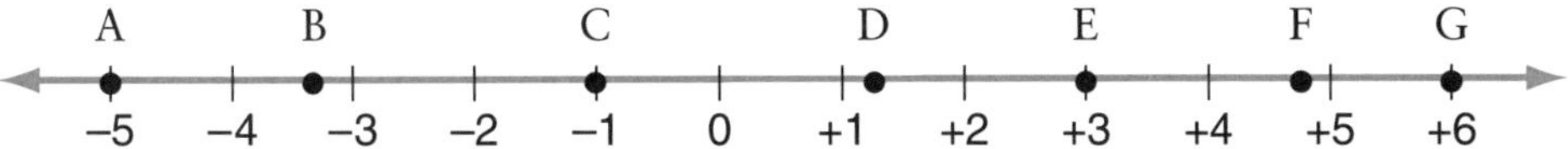

1. $+3$ E
2. -5 ______
3. $1\frac{1}{4}$ ______
4. 6 ______
5. $-3\frac{1}{3}$ ______
6. -1 ______
7. $4\frac{3}{4}$ ______

To check your answers, turn to page 195.

Adding Signed Numbers

Marty owes his brother Skip \$30, and he owes his brother Josh \$50. How much does Marty owe altogether?

A debt is like a negative number. The total of two debts is the sum of two negative numbers.

> To add numbers with the *same* signs, *add* and give the answer the sign of the numbers.

Notice how the problems in the examples are written. In this example, the + sign between parentheses means to add.

Example: $(-30) + (-50) =$

Since the signs are the same, add and make the answer negative. $(^-30) + (^-50) = {}^-80$

➠ Altogether Marty owes \$80.

When you add a negative number and a positive number, you *compare* the numbers. A number is *greater* than any number to its left on the number line. For example, +5 is greater than −8. However, −8 is farther from zero than +5. The number −8 has a greater **absolute value** than +5. The absolute value of a number is its distance from zero. The absolute value of +5 is 5. The absolute value of−8 is 8.

To add numbers with *different* signs, *subtract.* Then give the answer the sign of the number with the greater absolute value.

Example: $-7 + 13 =$

Since the signs are different, subtract. Since 13 has the greater absolute value, make the answer positive. The answer is +6.

$-7 + 13 = +6$

Remember, a number without a sign is positive.

Example: $(+8) + (-12) =$

Since the signs are different, subtract. Since −12 has the greater absolute value make the answer negative. The answer is −4.

$(+8) + (-12) = -4$

PRACTICE 63

A. Solve each problem.

1. $-6 + 9 =$ $(+3) + (-5) =$ $(-8) + (+9) =$

2. $-17 + 17 =$ $(-20) + (-19) =$ $+7 + 8 =$

3. $(-13) + (-12) =$ $+35 + 14 =$ $-18 + 21 =$

To add more than two signed numbers:

1. Add the positive numbers and make the total positive.
2. Add the negative numbers and make the total negative.
3. Subtract the two totals and give the answer the sign of the total with the greater absolute value.

Example: $+6 - 9 - 4 + 10 =$

Step 1.	Add the positive numbers.	$+6 + 10 = +16$
Step 2.	Add the negative numbers.	$-9 - 4 = -13$
Step 3.	Subtract the totals. Since +16 has the greater absolute value, make the answer positive. The answer is +3.	$+16 - 13 = +3$

B. Solve each problem.

4. $+9 - 10 + 4 =$ $\qquad (-3) + (+7) + (-8) + (12) =$

5. $-12 + 4 - 3 + 7 =$ $\qquad (-15) + (+12) + (-2) + (+6) =$

To check your answers, turn to page 196.

Subtracting Signed Numbers

On her June checking account statement, Shirley had a balance of $80. On her July statement, she had an overdraft of $30. Find the total of the checks that Shirley wrote from June to July.

A balance of $80 is positive. An overdraft of $30 is negative. To find the total Shirley spent, find the difference between +$80 and −$30.

> To subtract signed numbers, change the sign of the number being subtracted and the subtraction sign to an addition sign. Then add the signed numbers.

Example: $(+80) - (-30) =$

Step 1.	Change the sign of the number being subtracted, −30 to +30, and the subtraction sign to addition.	− becomes + $(+80) - (-30) = (+80) + (+30)$ subtraction becomes addition
Step 2.	Follow the rules for addition.	$(+80) + (+30) = +110$

➡ Shirley wrote checks for a total of $110 from June to July.

Example: $(+18) - (+7) =$

STEP 1.	Change the sign of the number being subtracted, +7 to −7, and the subtraction sign to addition.	+ becomes − $(+18) - (+7) = (+18) + (-7)$ subtraction becomes addition
STEP 2.	Follow the rules for addition. The answer is +11.	$(+18) + (-7) = +11$

PRACTICE 64

Solve each problem.

1. $(-6) - (-10) =$ $\quad$ $5 - (+12) =$ $\quad$ $-13 - (-9) =$

2. $(-25) - (30) =$ $\quad$ $(-21) - (-8) =$ $\quad$ $32 - (+12) =$

To check your answers, turn to page 196.

Multiplying Signed Numbers

Bill is taking his diet seriously. He plans to lose three pounds a week. What will be his weight in four weeks compared to now?

Think of a loss of weight as a minus (−) and a gain of weight as a plus (+). Think of time in the future as plus (+) and time in the past as minus (−).

When you multiply two signed numbers:

1. The answer is positive if the signs are alike.
2. The answer is negative if the signs are different.

Example: $(-3)(4) =$

Multiply and make the answer negative since the signs are different. $\quad$ $(-3)(4) = -12$

➡ In four weeks, Bill will weigh twelve pounds less (−12) than now.

Remember that besides the times sign ($\times$), multiplication can be shown:

1. with a raised dot ($\cdot$), or $5 \cdot 2$
2. with no sign before or between parentheses. $5(2)$ or $(5)(2)$

Example: $-6 \cdot -8 =$

Multiply and make the answer positive since the signs are alike. $-6 \cdot -8 = +48$

➧ The answer is +48 or 48.

PRACTICE 65

Solve each problem.

1. $9(-3) =$	$-12(7) =$	$-4(-8) =$
2. $+5(-11) =$	$3(+16) =$	$-10(-9) =$
3. $-2 \cdot 13 =$	$+15 \cdot +4 =$	$-1 \cdot -23 =$
4. $4 \cdot -20 =$	$+16 \cdot -3 =$	$-8 \cdot 25 =$
5. $(-7)(-9) =$	$(-8)(0) =$	$(36)(-3) =$

Use the following information to answer questions 6–8.

Think of a loss of weight as a minus (−) and a gain of weight as a plus (+). Think of time in the future as plus (+) and time in the past as minus (−).

6. Fred has been gaining three pounds a week. What will be his weight in four weeks compared to now?

7. Linda has also been gaining three pounds a week. What was her weight four weeks ago compared to now?

8. Melba has been losing three pounds a week. What was her weight four weeks ago compared to now?

To check your answers, turn to page 196.

Dividing Signed Numbers

The rules for dividing signed numbers are similar to the rules for multiplying signed numbers.

When you divide signed numbers:

1. The answer is positive if the signs are alike.
2. The answer is negative if the signs are different.

Example: $\frac{-30}{5} =$

Divide and make the answer negative since the signs are different. $\frac{-30}{5} = -6$

➡ The answer is −6.

Example: $\frac{-24}{-8} =$

Divide and make the answer positive since the signs are alike. $\frac{-24}{-8} = +3$

➡ The answer is +3 or 3.

If the divisor (bottom number) is larger than the top, reduce.

Example: $\frac{-6}{8} =$

Reduce and make the answer negative since the signs are different. $\frac{-6}{8} = \frac{-3}{4}$

To review reducing, turn to page 54.

➡ The answer is $\frac{-3}{4}$.

PRACTICE 66

Solve each problem.

1. $\frac{35}{-7} =$ $\quad \frac{-18}{-3} =$ $\quad \frac{-27}{9} =$

2. $\frac{-12}{-24} =$ $\quad \frac{-250}{-25} =$ $\quad \frac{-15}{35} =$

3. $\frac{-72}{-8} =$ $\quad \frac{16}{-24} =$ $\quad \frac{-10}{-10} =$

To check your answers, turn to page 196.

Evaluating Expressions

An **expression** is a mathematical operation or instruction written with numbers and symbols. To **evaluate** means to find the value.

Mathematicians have agreed on the following order for performing operations when an expression calls for more than one operation.

Order of Operations:

1. Parentheses and division bars
2. Powers and roots
3. Multiplication and division from left to right
4. Addition and subtraction from left to right

Example: Find the value of $9 + 2 \cdot 3$.

STEP 1. The expression has addition and multiplication. Since multiplication comes before addition in the order of operations, multiply first.

$9 + 2 \cdot 3 =$
$9 + 6 =$

STEP 2. Add 9 and 6.

$9 + 6 = 15$

➡ $9 + 2 \cdot 3 = 15$

Example: Find the value of $4 \cdot 5^2$.

Step 1. The expression has multiplication and power. Since powers come before multiplication in order of operations, find 5^2 first.

$4 \cdot 5^2 =$
$4 \cdot 25 =$

Step 2. Multiply 4 times 25.

$4 \cdot 25 = 100$

➠ $4 \cdot 5^2 = 100$

PRACTICE 67

A. Evaluate each expression.

1. $20 - 3 \cdot 5 =$ $\qquad$ $3(5) + 4(9) =$ $\qquad$ $12 - \frac{21}{3} =$

2. $9 \cdot 6 - 5 \cdot 4 =$ $\qquad$ $\frac{50}{2} + 3(10) =$ $\qquad$ $12 + 6^2 =$

Example: Evaluate the expression $\frac{15 - 7}{2}$.

Step 1. This expression has a division bar. It groups $15 - 7$. Do the operation grouped by the division bar first. Subtract $15 - 7$.

$\frac{15 - 7}{2} = \frac{8}{2}$

Step 2. Divide 8 by 2.

$\frac{8}{2} = 4$

➠ $\frac{15 - 7}{2} = 4$

B. Evaluate each expression.

3. $2(14 - 3) =$ $\qquad$ $\frac{9 + 6}{3} =$ $\qquad$ $\frac{12}{1 + 5} =$

4. $(7 - 2)^2 =$ $\qquad$ $\frac{13 + 8}{7} =$ $\qquad$ $\frac{1}{2}(17 - 3) =$

Many algebraic expressions contain letters. These letters, also called *variables* or *unknowns,* are used in place of numbers. To evaluate an expression containing a letter, you must **substitute** a number for the letter.

In the next problems, you will see yet another way to write multiplication in algebra. The expression $5a$ means "five multiplied by some unknown a." The expression mn means "some unknown m multiplied by some other unknown n."

Example: Find the value of $5a$ when $a = 7$.

STEP 1. Substitute 7 for a in the expression. $5a = 5 \cdot 7$

STEP 2. Multiply 5 times 7. $5 \cdot 7 = 35$

➡ When $a = 7$, $5a = 35$.

C. Evaluate each expression. Use the order of operations.

5. $7 + x$ when $x = 3$ $c - d$ when $c = 12$ and $d = 5$

6. $2(x + 4)$ when $x = 8$ $3a + 4b$ when $a = 6$ and $b = 2$

To check your answers, turn to page 197.

Writing Expressions

Jamal doesn't like to brag, but he says that he makes twice what he made when he started working.

Algebra is a useful tool for expressing number relationships.

Example: When Jamal started working, he made s dollars a year. Now he makes twice as much. Write an expression for his salary now.

STEP 1. Look for a word that suggests an operation. The word *twice* suggests multiplying by 2.

STEP 2. Write 2 next to his starting salary s. $2s$

➡ Jamal's salary now can be represented by $2s$.

PRACTICE 68

A. Write an algebraic expression for each.
For problems 1–3, use *n* to represent each unknown number.

1. three times a number a number increased by ten

2. nine less than a number a number divided by five

3. the number decreased by two a number squared

For problems 4–10, use the given unknown to write an expression.

4. twenty more than r seven less than z

5. the sum of eight and w t divided by fifteen

6. Montiel makes p dollars a month. He tries to save one-tenth of his income. Write an expression for the amount he saves each month.

7. Helen is y years old. Her daughter Katie is 23 years younger. Write an expression for Katie's age.

8. Eddie makes w dollars a week. He works 35 hours every week. Write an expression for the amount he makes each hour.

9. Gloria's gross salary is g dollars. Her employer deducts 21% of her gross salary taxes. Write an expression for the amount of the deductions. [**Hint:** Change the percent to a decimal.]

Read multi-step operations carefully.

Example: Write an expression for six times the sum of five and a number n.

STEP 1. Since six is multiplied by a sum, first group the sum in parentheses. $(5 + n)$

STEP 2. Then write 6 outside the parentheses. $6(5 + n)$

B. Write an algebraic expression for each.

10. one-third of the sum of x and 9

11. twice the difference of n and 10

12. half the difference of z and seven

13. ten divided into the sum of s and three

To check your answers, turn to page 197.

Using Formulas

Michelle drove at a rate of 55 mph for 3 hours. How far did she drive?

A **formula** is a mathematical instruction written in the language of algebra. The distance formula $d = rt$ means "distance equals rate times time," where

d is *distance,* usually measured in miles,
r is *rate,* usually measured in miles per hour (mph), and
t is *time,* usually measured in hours.

Example: Find the distance Michelle drove in 3 hours at 55 mph.

STEP 1. Substitute 55 for r and 3 for t in the formula $d = rt$.

$d = rt$
$d = 55 \cdot 3$

STEP 2. Multiply 55 times 3. $d = 165$ miles

➡ Michelle drove 165 miles.

PRACTICE 69

Use the distance formula to solve these problems.

1. Kareem drove for 2.5 hours at an average speed of 60 mph. How far did he drive?

2. Deborah jogs at an average of 6.5 mph. How far can she jog in two hours?

3. A plane flew at an average speed of 418 mph. How far did it travel in 3.5 hours?

4. Michelle is driving across the country. On the first day of her trip, she plans to drive six hours. If she averages 65 miles per hour, how far will she drive the first day?

5. Driving a truck through a mountainous area, Stan estimates he can drive at an average rate of 30 miles per hour. If he drives for $4\frac{1}{2}$ hours at that rate, how many miles will he travel?

To check your answers, turn to page 198.

Understanding Equations

An equation is a statement that two amounts are equal. Compare these examples.

Numbers and Symbols	*Words*
$5 + 4 = 9$	five plus four is equal to nine
$7c = 35$	seven times *c* is equal to thirty-five
$20 = m - 8$	twenty is equal to *m* minus 8
$\frac{s}{6} = 10$	*s* divided by six is equal to ten

PRACTICE 70

For each equation written in words, choose the corresponding equation written in numbers and symbols.

1. The sum of twelve and x is nineteen.

   a. $12x$ *b.* $12 + x$ *c.* $12 + x = 19$ *d.* $12x = 19$

2. The number n decreased by seven is equal to fifteen.

   a. $n - 7 = 15$ *b.* $\frac{n}{7} = 15$ *c.* $15n = 7$ *d.* $7n = 15$

3. Thirty is equal to r divided by two.

   a. $30 - r = 2$ *b.* $30r = 2$ *c.* $\frac{30}{2} = r$ *d.* $30 = \frac{r}{2}$

4. Eight times y is equal to fifty.

   a. $8 + y = 50$ *b.* $8y = 50$ *c.* $8 - y = 50$ *d.* $\frac{8}{y} = 50$

For each word equation, write an equation in numbers and symbols.

5. The number s decreased by thirteen is equal to twenty-one.

6. Nine is equal to two times z.

7. The sum of p and six is equal to fifty.

8. The product of x and twelve is equal to one hundred eight.

To check your answers, turn to page 198.

One-Step Equations

Find the solution to the equation $x + 7 = 23$.

The **solution** is the value of the unknown that makes an equation a true statement. It is easy to find the solution to some equations. For $x + 7 = 23$, you know that $16 + 7 = 23$. The solution for x is 16. Not all equations are so easy.

To solve equations, use opposite or **inverse** operations.
Subtraction is the inverse of addition.
Addition is the inverse of subtraction.
Division is the inverse of multiplication.
Multiplication is the inverse of division.

To solve a one-step equation, do the inverse operation on both sides of the equation. You often see the word *side* in instruction about equations. The = sign separates an equation into two sides.

Example: Solve $x + 7 = 23$.

STEP 1.	Identify the operation in the equation. Since 7 is added to the unknown, the operation is addition.	addition
STEP 2.	Do the inverse operation. Subtract 7 from both sides.	$x + 7 = 23$ $x + 7 - 7 = 23 - 7$
STEP 3.	Simplify both sides.	$x = 16$

➡ The solution is $x = 16$.

It is often easy to find the solution to one-step equations. However, this is the time to develop skills that will make solving longer equations easier.

Example: Solve $3n = 45$.

STEP 1.	Identify the operation in the equation. Since the unknown is multiplied by 3, the operation is multiplication.	multiplication

Step 2. Do the inverse operation, division, to both sides. Divide both sides by 3. $\frac{3n}{3} = \frac{45}{3}$

When you simplify $\frac{3n}{3}$, you get 1n. This is the same as n.

Step 3. Simplify both sides. $n = 15$

➠ The solution is $n = 15$.

PRACTICE 71

Solve each equation.

1. $9a = 72$ $\quad 41 = b - 12$ $\quad c + 1.5 = 6$

2. $\frac{d}{13} = 2$ $\quad 6 = 8e$ $\quad f - 6 = 19$

3. $g + 12 = 200$ $\quad 16 = \frac{h}{5}$ $\quad 20k = 300$

4. $9 = 18s$ $\quad t - 2 = 8$ $\quad u + 11 = 7$

Write and solve an equation for each problem. Use *n* for each unknown.

5. If seven is added to a number, the difference is nineteen. What is the number?

6. Sixteen is equal to a number divided by four. Find the number.

7. Tiffany makes $22 dollars an hour for over-time work. This is twice her regular wage. Find her regular wage.

8. Carlos saved $28 buying a sweater on sale. The sale price was $35. Find the original price of the sweater.

To check your answers, turn to page 198.

Two-Step Equations

Jonelle's age is five more than twice her son's age. If Jonelle is 37, how old is her son?

An equation for their ages is $2s + 5 = 37$ where s represents Jonelle's son's age.

To solve two-step equations, use inverse operations in the following order:

1. Do addition or subtraction operations first.
2. Do multiplication or division operations last.

Example: Solve $2s + 5 = 37$.

Step 1. Do the inverse of addition first. Subtract 5 from both sides.

$$2s + 5 - 5 = 37 - 5$$
$$2s = 32$$

Step 2. Do the inverse of multiplication. Divide both sides by 2.

$$\frac{2s}{2} = \frac{32}{2}$$
$$s = 16$$

➡ Jonelle's son is 16 years old.

Example: Solve $3 = \frac{m}{4} - 9$.

Step 1. Do the inverse of subtraction first.

Add 9 to both sides.

$$3 + 9 = \frac{m}{4} - 9 + 9$$
$$12 = \frac{m}{4}$$

Step 2. Do the inverse of division. Multiply both sides by 4.

$$12 \cdot 4 = \frac{m}{4} \cdot 4$$
$$48 = m$$

➡ The solution is $m = 48$.

PRACTICE 72

Solve each problem.

1. $8m + 7 = 55$	$5a - 3 = 17$	$\frac{x}{6} + 3 = 7$
2. $37 = 4c + 9$	$47 = 10x - 3$	$1 = \frac{m}{9} - 4$
3. $3a - 7 = 23$	$\frac{1}{2}x + 5 = 12$	$\frac{n}{15} - 1 = 3$
4. $3a + 7 = -5$	$13 = \frac{c}{4} + 5$	$2m - 9 = 1$

Write and solve an equation for each problem. Use *m* for each unknown.

5. Five less than twice a number is nine. Find the number.

6. Ten more than three times a number is twenty-two. What is the number?

7. Four less than one-half of a number is seven. Find the number.

8. From September to October, Dave's oil delivery service gets busier. In October he made five more than four times as many deliveries as he made in September. Dave made 65 deliveries in October. How many did he make in September?

9. Joaquin is a plumber. In an hour he makes $3 less than twice as much as his assistant. Joaquin makes $37 an hour. How much does his assistant make?

To check your answers, turn to page 199.

Equations with Separated Unknowns

Sergio worked 6 hours on Monday and 3 hours on Tuesday pouring a foundation. He made \$144. How much did he make each hour?

The equation $6h + 3h = 144$ expresses the problem. When the unknowns are separated, combine them according to the rules addition of signed numbers. 6h and 3h are called *like terms*. Like terms have the same variable.

Example: Solve $6h + 3h = 144$.

STEP 1. Combine the like terms.

$$6h + 3h = 144$$
$$9h = 144$$

STEP 2. Divide both sides by 9.

$$\frac{9h}{9} = \frac{144}{9}$$
$$h = 16$$

➡ Sergio made \$16 an hour.

When combining terms, remember that an unknown such as m is the same as $1m$.

Example: Simplify the expression $4x + 8 + x - 6$.

STEP 1. Add the terms with x. $4x + x = 5x$

STEP 2. Subtract 6 from 8. $8 - 6 = 2$

$$4x + 8 + x - 6 = 5x + 2$$

PRACTICE 73

A. Simplify each expression.

1. $9m - 3 + 2m - 7 =$ $\qquad$ $3a - a + 8 =$

2. $12 - 7y - 6 + -5y =$ $\qquad$ $13 + c - 6 + 12c =$

Solve each equation.

3. $5a - a = 24$ $\qquad$ $15 = 4x + x$

4. $18c + 7c = 50$ $\qquad$ $36 = 11c - 2c$

5. $6y - 7 - 5y = 14$ $\qquad$ $9r - 2r + 3 = 31$

When the variables in an equation are on opposite sides of the = sign, combine them using inverse operations.

Example: Solve for x in $7x = 2x + 20$.

STEP 1. To combine the terms with x, subtract 2x from both sides.

$$7x = 2x + 20$$
$$7x - 2x = 2x - 2x + 20$$

STEP 2. Divide both sides by 5.

$$\frac{5x}{5} = \frac{20}{5}$$
$$x = 4$$

B. Solve each equation.

6. $8e = 30 + 5e$ $\qquad$ $3h = 12 + 2h$

7. $16 - 2m = 6m$ $\qquad$ $6z = 10 + z$

8. $5y + 8 = 3y + 26$ $\qquad$ $8n - 9 = 7n + 13$

9. Ten times the number n equals three times the same number n increased by seven. Find the number.

10. Seven times a number n increased by twelve equals twice the number n decreased by eight. Find the number.

To check your answers, turn to page 200.

Equations with Parentheses

Twice the sum of x and five equals sixteen. Find x.

The equation $2(x + 5) = 16$ expresses the problem. Before you can solve an equation with parentheses, you need to multiply to get rid of the parentheses. In this equation, 2 is multiplied by both x and 5.

Example: Solve $2(x + 5) = 16$.

STEP 1. Multiply to get rid of the parentheses. Multiply 2 times x and 2 times 5.

$$2(x + 5) = 2x + 10$$
$$2x + 10 = 16$$

STEP 2. Subtract 10 from both sides.

$$2x + 10 - 10 = 16 - 10$$

STEP 3. Divide both sides by 2.

$$\frac{2x}{2} = \frac{16}{2}$$
$$x = 8$$

➡ $x = 8$

PRACTICE 74

Solve each equation.

1. $3(a + 2) = 18$ $\quad$ $5(n - 3) = 20$

2. $6(y - 5) = 42$ $\quad$ $60 = 3(m + 4)$

3. $3(r + 4) = 2r + 17$ $\quad$ $9(c - 2) = c + 30$

4. Six times the difference of n and two equals eighteen. Find the value of n.

5. Seven times the sum of n and five equals fifty-six. Find the value of n.

To check your answers, turn to page 201.

Using Formulas Like Equations

Miranda drove 108 miles in two hours. What was her rate of speed?

You have used the distance formula $d = rt$ to calculate distance. You can also use the formula to calculate the rate when you know both distance and time. If you know any two variables, you can solve the distance formula like an equation to find the third variable.

Example: Find Miranda's rate of speed if she went 108 miles in 2 hours.

STEP 1. Substitute 108 for d and 2 for t in the distance formula.

$$d = rt$$
$$108 = r \cdot 2$$

STEP 2. Divide both sides by 2.

$$\frac{108}{2} = r \cdot \frac{2}{2}$$
$$54 = r$$

➡ Miranda's rate was 54 mph.

PRACTICE 75

Use the distance formula $d = rt$ to solve problems 1–4.

1. Keith drove 160 miles in 2.5 hours. What was his average driving speed?

2. On the highway, Phyllis sticks to the speed limit of 65 miles per hour. How long will it taker her to reach a destination 260 miles away?

3. Chris can bicycle at 16 mph. How long will he need to travel 40 miles?

4. The distance between two cities is 360 miles. A train took 4.5 hours to travel between the cities. What was the average speed of the train?

The formula $c = nr$ is the formula for finding total cost when you know the number (n) you are buying, and the rate (r), the cost for one. Use the formula to solve problems 5–8.

5. Frank paid $7.20 for 2.5 pounds of beef. What was the price per pound (the rate)?

6. Myrna paid $38.85 for three gallons of paint. What was the cost of one gallon?

7. Sliced ham was priced at $2.90 a pound. Emma paid $4.35 for the amount of ham she needed. Find the weight of the ham she bought.

8. Marlen paid $19.20 for 12 feet of lumber. What was the price per foot?

The formula $m = \frac{1}{3}(a + b + c)$ is the formula for finding the average (mean) for three items where *a*, *b*, and *c* are the items. Use the formula to solve problems 9–10.

9. On his first math quiz, Alex got a score of 78. On the second quiz, his score was 84. What score does he have to get on the third quiz to have an average score of 85?

10. Marta's June phone bill was $36. Her July phone bill was $42. Her average monthly phone bill for June, July, and August was $44. Find the amount of her August bill.

To check your answers, turn to page 201.

ALGEBRA REVIEW

These problems will help you find out if you need to review the algebra section of this book. Solve each problem. When you finish, look at the chart to see which pages you should review.

1. $13^2 =$

2. $9^2 - 6^2 =$

3. $\sqrt{49} + \sqrt{64} =$

For problems 4–7, use the number line below to tell the letter that represents each of the following values.

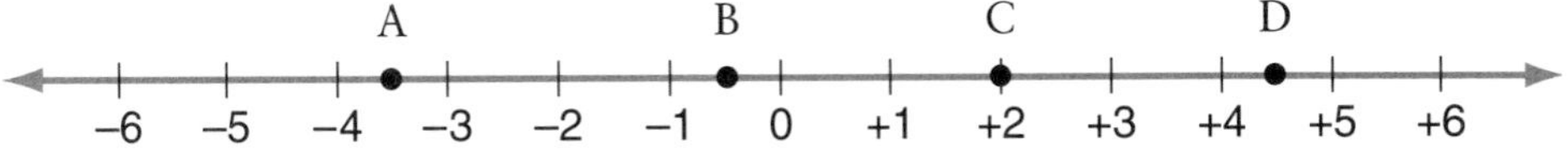

4. $-3\frac{1}{2}$

5. 2

6. $-.5$

7. $4\frac{1}{2}$

For problems 8–15, find the value of each.

8. $-17 + 24 =$

9. $15 - 9 - 11 + 3 =$

10. $8 - (-13) =$

11. $+11 \cdot -6 =$

12. $(-3)(-11) =$

13. $\frac{56}{-8} =$

14. $5(12 - 7) =$

15. $8 \cdot 9 - \frac{20}{4} =$

16. What is the value of $3(x + 7)$ when $x = -2$?

17. Find the value of $4m - 2n$ when $m = 7$ and $n = 6$.

For problems 18–19, write an algebraic expression for each verbal expression.

18. twenty less than n

19. twice the sum of s and twelve

For problems 20–25, solve each equation.

20. $40 = x - 17$

21. $\frac{m}{9} = 12$

22. $7c - 6 = 50$

23. Five less than half a number is nine. Find the number.

24. $9d - 5d = 100$

25. $5(m + 2) = 40$

26. Jorge drove his truck 192 miles in four hours. What was his average driving speed?

PROGRESS CHECK

Check your answers on page 202. Then return to the review pages for the problems you missed. Correct your answers before going on to the next unit.

If you missed problems	*Review pages*
1 to 3	115 to 117
4 to 7	118
8 to 15	119 to 124
16 to 19	125 to 128
20 to 26	129 to 140

UNIT 7

Geometry

Lines and Angles

Geometry is a branch of mathematics that deals with the measurement of lines, angles, surfaces, and three-dimensional figures. Geometry has many special terms, some of which you probably already know. In this unit, try to memorize the terms you do not know.

A line that runs straight up and down is called a **vertical** line. A telephone pole is an example. A line that runs from left to right is a **horizontal** line. The edge of a tabletop is an example.

vertical

horizontal

Two lines that cross each other are called **intersecting** lines. The symbol at railroad crossings is an example.

Lines that are the same distance from each other are **parallel** lines. Railroad tracks are an example.

Lines that meet to form square corners (also called *right angles*) are called **perpendicular** lines. The top edge and side of this book are examples.

PRACTICE 76

A. Fill in each blank with a word that correctly completes each sentence.

1. Lines A and B are called ______________ lines.

2. Lines C and D are called ______________ lines.

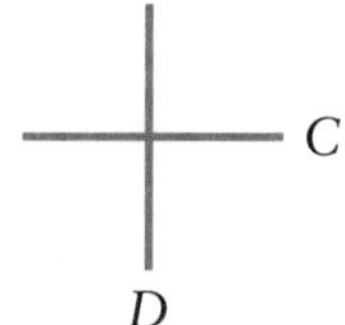

An **angle** is formed by two lines or **rays** that extend from the same point. The two lines or rays form the **sides** of an angle. The point where the sides meet is called the **vertex.** The symbol ∠ means *angle.*

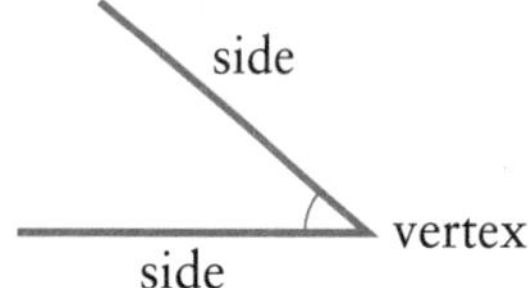

The size of an angle depends on how open or closed the sides are. ∠*a* is smaller than ∠*b*.

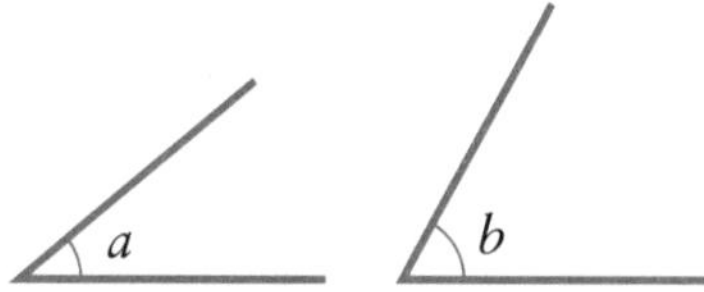

There are four kinds of angles.

A **right angle** has 90°. The sides of a right angle are *perpendicular.* A small square at the vertex indicates a right angle. ∠*X* is a right angle.

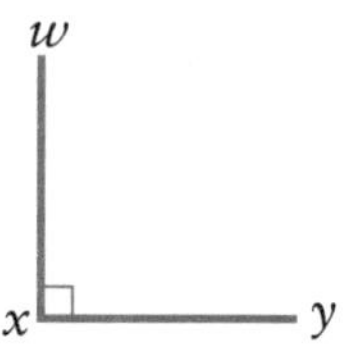

An **acute angle** has less than 90°. The sides of an acute angle are "closed" more than the sides of a right angle. ∠*DEF* is an acute angle.

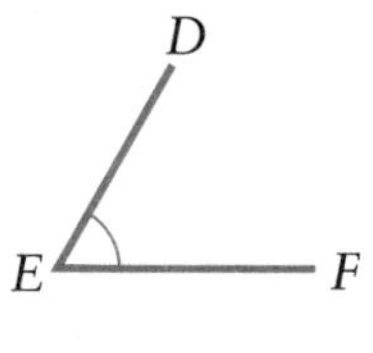

An **obtuse angle** has more than 90° and less than 180°. The sides of an obtuse angle are "open" more than the sides of a right angle. ∠*c* is obtuse.

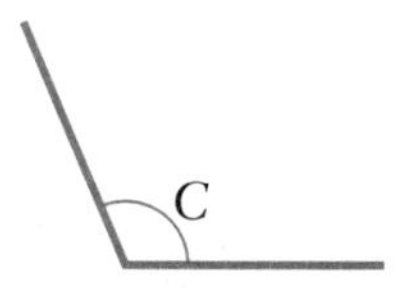

A **straight angle** has exactly 180°. A straight angle looks like a straight line. ∠*POQ* is straight.

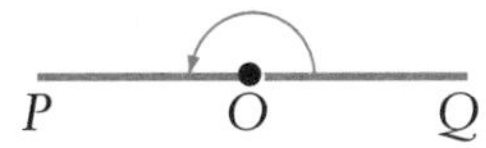

B. Identify each angle.

3. 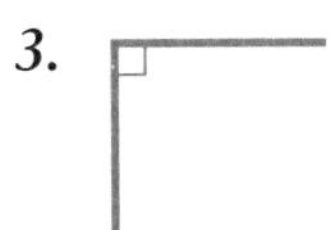

4. 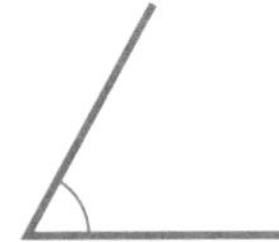

5. 

6. 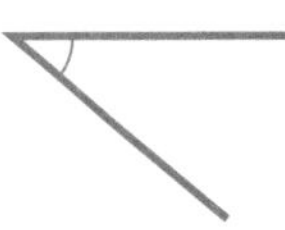

7. 40°

8. 140°

9. 90°

10. 180°

To check your answers, turn to page 203.

Pairs of Angles

What is the measurement of ∠*a* in the illustration?

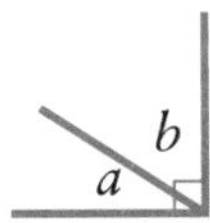

Together ∠*a* and ∠*b* form a right angle. Two angles whose sum is 90° are **complementary angles.** To find ∠*b*, subtract 32° from 90°.

Example: Find ∠*b* in the illustration above.

Subtract 32° from 90°. 90° − 32° = 58°

➡ ∠*x* = 58°.

Two angles whose sum is 180° are **supplementary angles.**

Example: ∠*AOB* = 47°. Find ∠*BOC*.

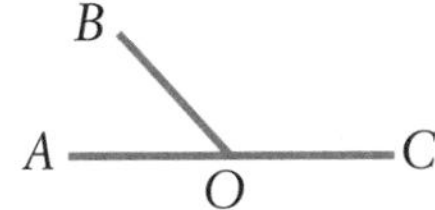

Subtract 47° from 180°. 180° − 47° = 133°

➡ ∠*BOC* = 133°.

PRACTICE 77

Solve each problem.

1. In the illustration, $\angle DOE = 29°$. Find $\angle COD$.

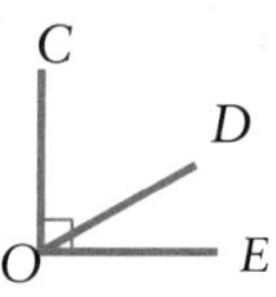

2. In the illustration, $\angle t = 117°$. Find $\angle s$.

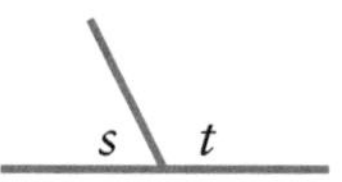

3. In the illustration, $\angle ABD = 125°$ and $\angle CBD = 55°$. What is the measurement of $\angle ABC$?

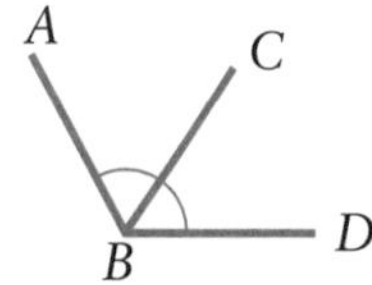

4. If an angle measures 74°, what is the measurement of its complement?

To check your answers, turn to page 203.

Common Geometric Figures

The three most common flat figures made up of straight sides are the rectangle, the square, and the triangle.

A **rectangle** has four sides and four right angles. Opposite sides (the sides across from each other) are equal. The longer side is called the *length*. The shorter side is called the *width*.

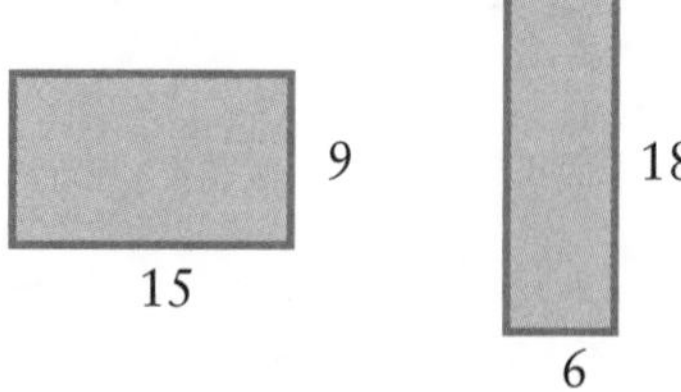

A **square** has four equal sides and four right angles.

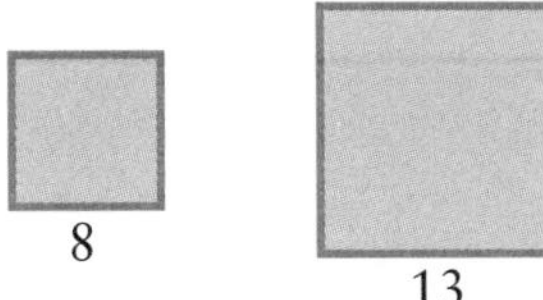

A **triangle** has three sides. The side that the triangle appears to rest on is called the *base*. The perpendicular distance from the base to the highest point is called the *height*.

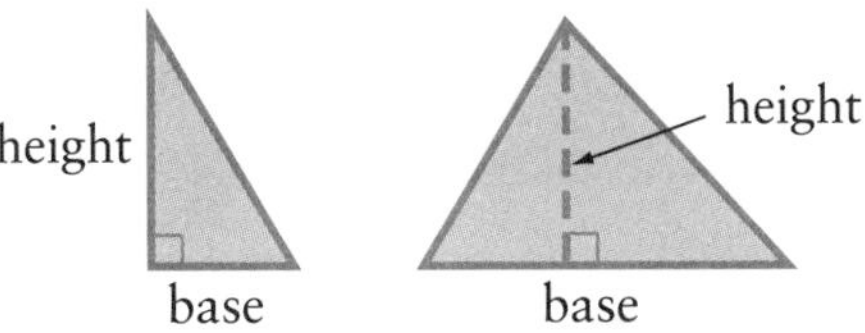

PRACTICE 78

A. Identify each shape.

1. 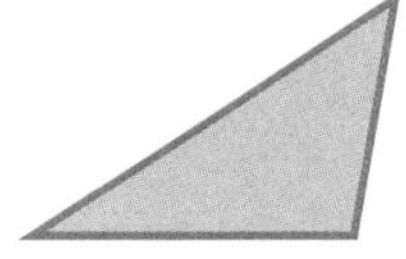

2. 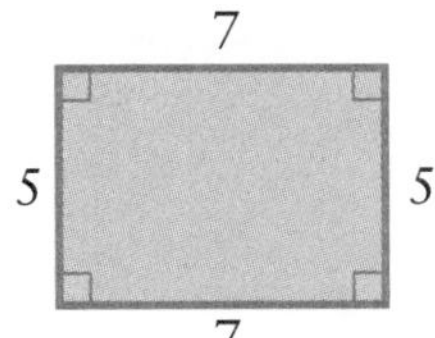

3. 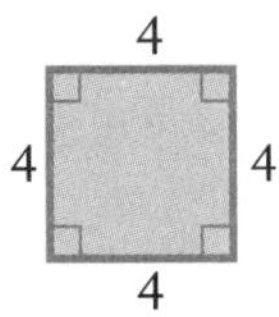

4. The window above Kendra's kitchen sink has four right angles and four sides each of which measures 42 inches. The window is what geometric shape?

5. In triangle *DEF*, what is the measurement of the height?

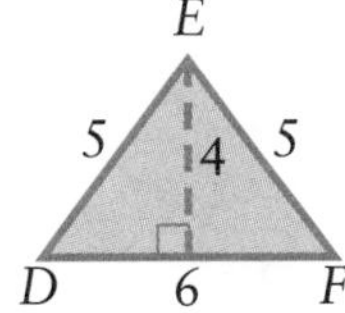

The other flat figure often studied in geometry is the **circle.** The distance around a circle is called the **circumference.** The distance from the center of a circle to any point on the circle is the **radius.** The length of a straight line that passes through the center of a circle is called the **diameter.**

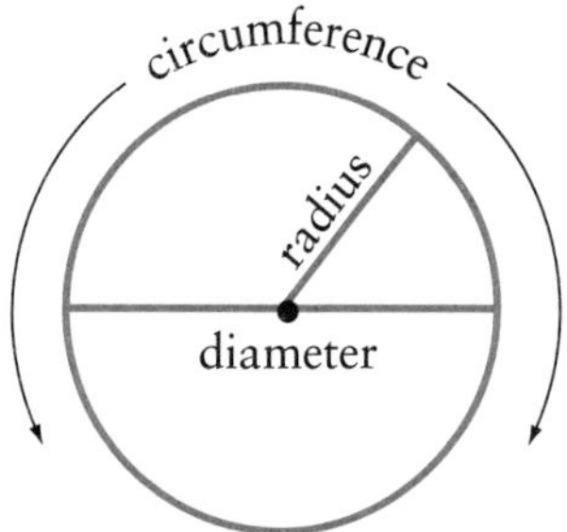

For any circle the measurement of the diameter is twice the radius.

B. Answer each question.

6. The distance around a circle is called the ______________.

7. Find the diameter of the circle shown here.

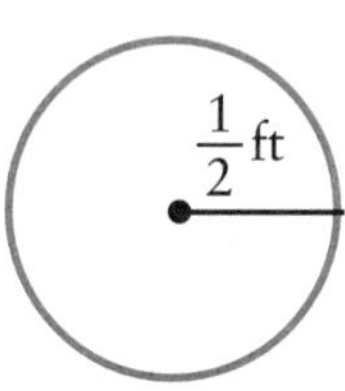

8. If r is the radius of a circle, which of the following represents the diameter?

   a. $\frac{r}{2}$ *b.* r^2 *c.* $r + 2$ *d.* $2r$

The three most common three-dimensional figures are the rectangular solid, the cube, and the cylinder.

A **rectangular solid** has length, width, and height (or depth). Each face or surface of a rectangular solid is either a rectangle or a square. Every corner is a right angle.

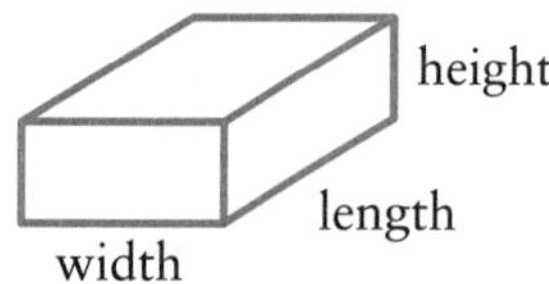

A **cube** has equal sides (or edges). Each face of a cube is a square.

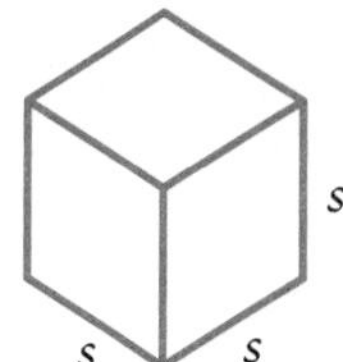

A **cylinder** has circles for its bases (top and bottom). The height of a cylinder is perpendicular to the bases.

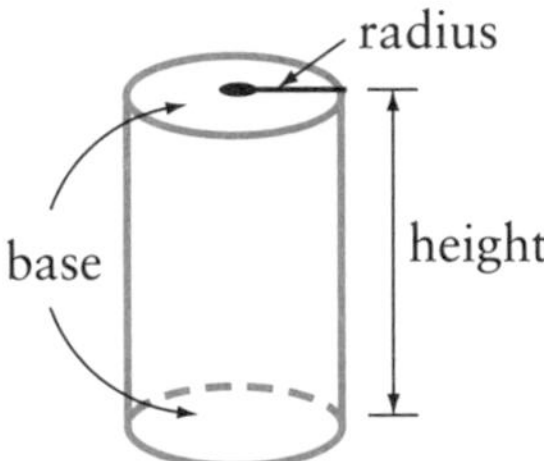

C. Answer each question.

9. In the rectangular solid pictured here, the width is half the length. Find the width.

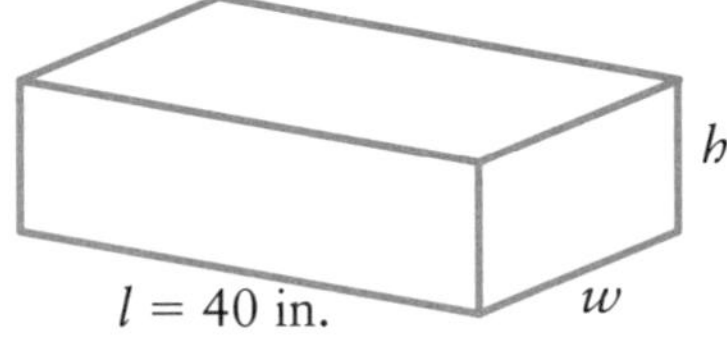

10. What is the diameter of the cylinder pictured here?

To check your answers, turn to page 203.

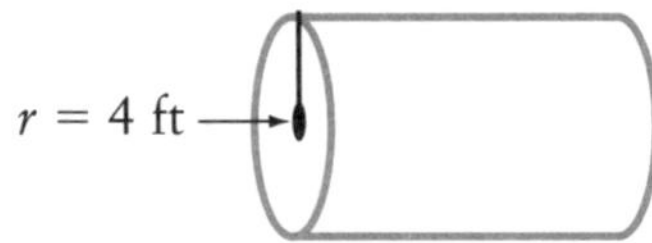

Perimeter and Circumference

Ralph repairs and sells used cars. He wants to fence in the space where he stores rebuilt cars. The diagram shows the layout of the space. How many feet of fencing does he need to enclose the space?

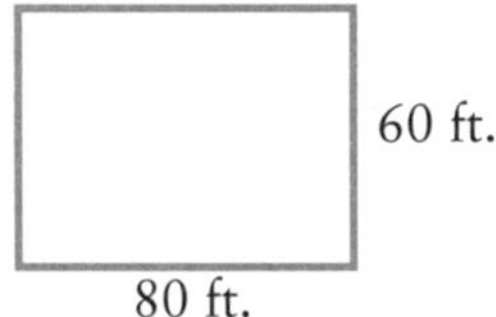

The distance around a flat figure is called the **perimeter.** Perimeter is measured in units such as yards, feet, inches, and meters.

To find the perimeter of a rectangle, use the formula $P = 2l + 2w$ where l is the length and w is the width.

Example: Find the perimeter of a rectangle 80 ft long and 60 ft wide.

STEP 1. Substitute 80 for l and 60 for w in the formula $P = 2l + 2w$.

$P = 2l + 2w$
$P = 2 \cdot 80 + 2 \cdot 60$
$P = 160 + 120$
$P = 280$ ft

STEP 2. Evaluate the formula.

➡ Ralph needs 280 feet of fencing.

PRACTICE 79

A. Use the formula $P = 2l + 2w$ to find the perimeter of each rectangle.

1. 7 m by 15 m

2. 20 yd. by 42 yd.

3. $3\frac{1}{2}$ in. by $7\frac{1}{2}$ in.

4. Gina built a rectangular stone terrace in her garden. The terrace is 24 feet long and 18 feet wide. She wants to put a brick boarder around the terrace. How many feet of bricks does she need?

To find the perimeter of a square, use the formula $P = 4s$ where s is the measurement of one side.

B. Use the formula $P = 4s$ to find the perimeter of each square.

5. 9 yd.

6. 16 mi

7. 6.2 m

8. Janet has 72 inches of picture frame molding. Assuming no waste, find the measurement of the longest side of a square picture that she can frame with the molding

To find the perimeter of a triangle, use the formula $\mathbf{P = a + b + c}$ where a, b, and c are the sides of a triangle.

C. Use the formula $P = a + b + c$ to find the perimeter of each figure.

9.

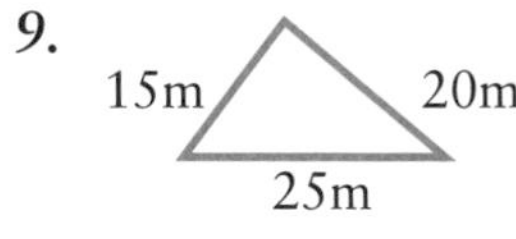

10.

16 ft.
16 ft.
16 ft.

11.

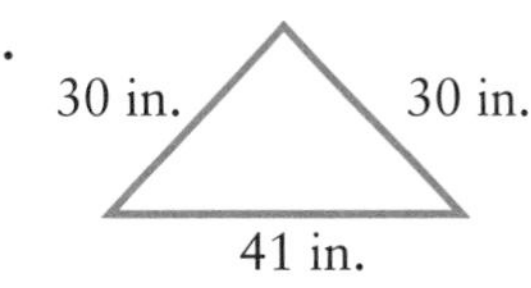

In formulas for circles, you will often see the Greek letter π (pi). This symbol represents the value of dividing the circumference of a circle by the diameter. The number is close to the mixed decimal 3.14.

You know that the distance around a circle is called the *circumference*. To find the circumference of a circle, use the formula $\mathbf{C = \pi d}$ where $\pi = 3.14$ and d is the diameter of the circle. You can also use the formula $\mathbf{C = 2\pi r}$ where $\pi = 3.14$ and r is the radius.

Example: What is the circumference of the circle illustrated here?

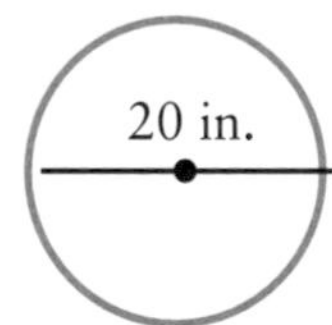

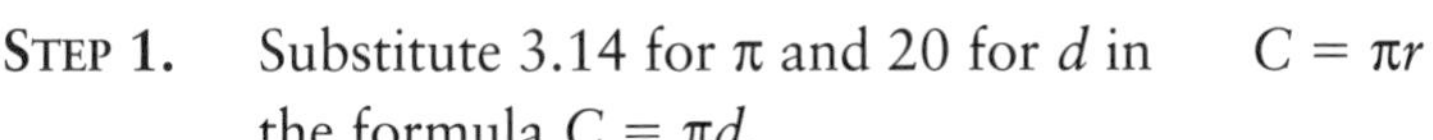

STEP 1. Substitute 3.14 for π and 20 for d in the formula $C = \pi d$. $\quad C = \pi r$

STEP 2. Evaluate the formula. $\quad C = 3.14 \cdot 20$

$C = 62.8$ in.

The circumference of the circle is 62.8 inches.

D. Use the formula $C = \pi d$ to find the circumference of each figure.

12.

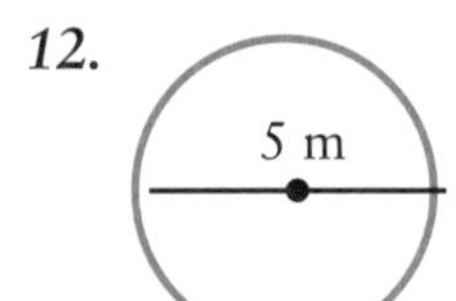

13.

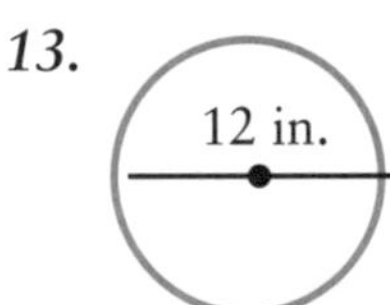

14.

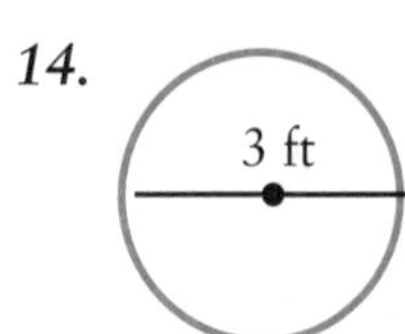

To check your answers, turn to page 203.

Area

What is the area of a garage floor that is 24 feet long and 20 feet wide?

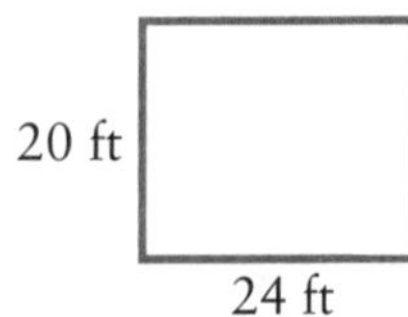

Area is a measure of the amount of *surface* on a flat figure. It is measured in square units such as square inches, square feet, and square meters. These units can be abbreviated with exponents. For example, ft^2 means square feet.

The rectangle pictured here is 4 inches long and 3 inches wide. It has a total of 12 square inches on its surface.

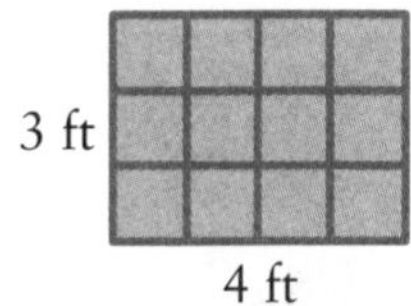

To find the area of a rectangle, use the formula $A = lw$, where l is the length and w is the width.

Example: Find the area of a rectangle 24 feet long and 20 feet wide.

STEP 1. Substitute 24 for l and 20 for w in the formula $A = lw$.

$A = lw$
$A = 24 \cdot 20$
$A = 480\ ft^2$

STEP 2. Evaluate the formula.

➡ The area of the garage floor is 480 square feet.

PRACTICE 80

A. Use the formula $A = lw$ to find the area of each rectangle.

1. 9 yd, 12 yd

2. 20 ft, 16 ft

3. 7.5 m, 10 m

4. The drawing shows the top of a glass-topped display case that Celia built for her store. What is the area of the top of the case?

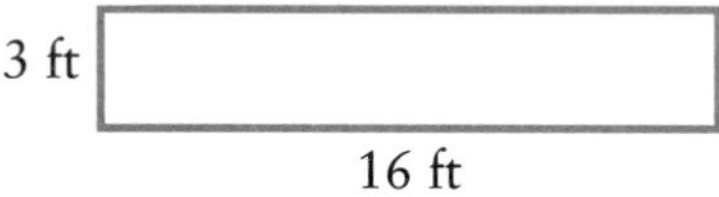

To find the area of a square, use the formula $\boldsymbol{A = s^2}$, where s is the side.

B. Use the formula $A = s^2$ to find the area of each square.

5. $s = 9$ ft

6. $s = 15$ in.

7.

$s = 1.2$ m

8. A square foot is a square measuring 12 inches on each side. How many square inches are in one square foot?

To find the area of a triangle, use the formula $\boldsymbol{A = \frac{1}{2}bh}$, where b is the base and h is the height.

C. Use the formula $A = \frac{1}{2}bh$ to find the area of each triangle.

9.

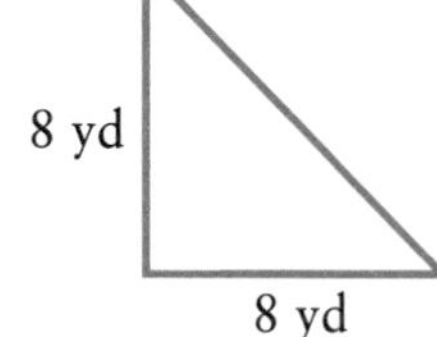

10.

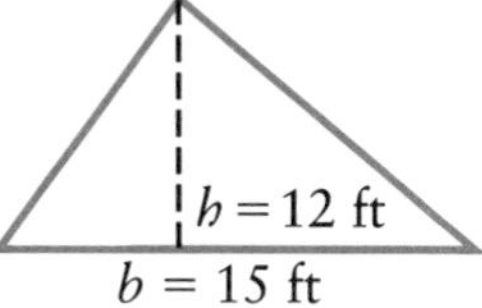

11.

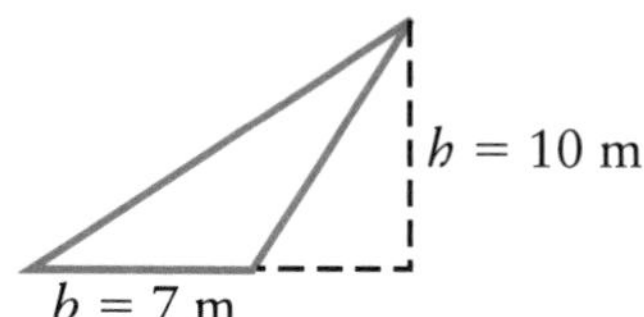

To find the area of a circle, use the formula $A = \pi r^2$, where π is 3.14 and r is the radius.

Example: Find the area of circle with a radius of 4 inches.

STEP 1.	Substitute 4 for r in the formula $A = \pi r^2$.	$A = \pi r^2$ $A = 3.14 \cdot 4^2$ $A = 3.14 \cdot 16$
STEP 2.	Evaluate the formula.	$A = 50.24 \text{ in.}^2$

➡ The area of the circle is 50.24 square inches.

D. Use the formula $A = \pi r^2$ to find the area of each circle.

12.

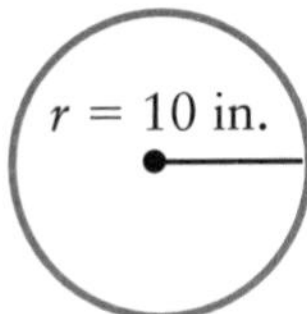

13.

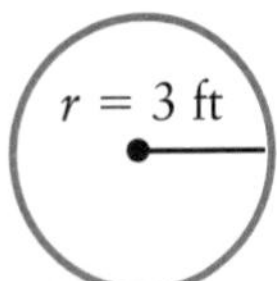

14.

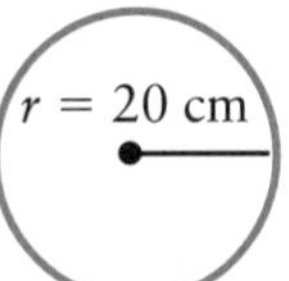

To check your answers, turn to page 204.

Volume

What is the capacity of the container pictured here?

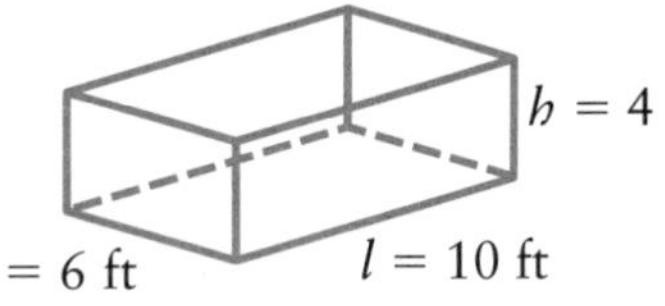

Capacity is another word for volume. **Volume** is a measure of the amount of space inside a three-dimensional figure. Volume is measured in cubic units such as cubic feet or cubic yards. These units can be abbreviated with exponents. For example, ft^3 means cubic feet.

To find the volume of a rectangular solid, use the formula $V = lwh$, where l is the length, w is the width, and h is the height.

Example: Find the volume of a rectangular container that is 10 ft long, 6 ft wide, and 4 ft long.

STEP 1.	Substitute 10 for l, 6 for w, and 4 for h in the formula $V = lwh$.	$V = lwh$ $V = 10 \cdot 6 \cdot 4$ $V = 240 \text{ ft}^3$
STEP 2.	Evaluate the formula	

➡ The capacity of the container is 240 cubic feet.

PRACTICE 81

A. Use the formula $V = lwh$ to find the volume of each figure.

1. 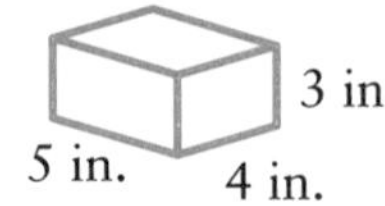

2. 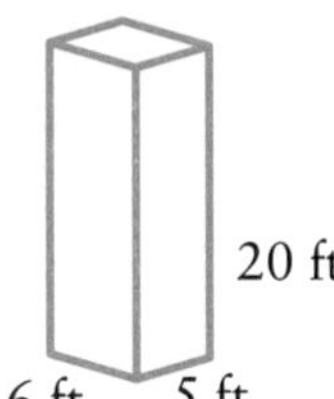

3. 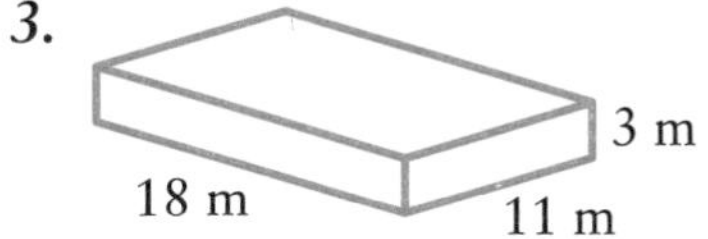

4. The cardboard carton shown in the illustration has a capacity of 3,840 cubic inches. What is the height of the container?

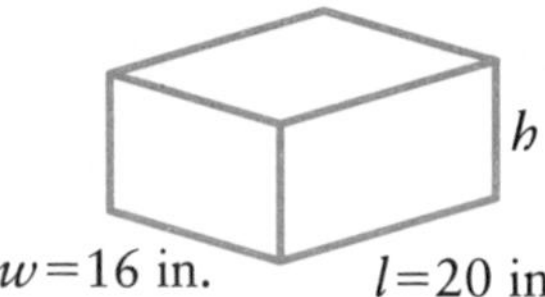

To find the volume of a cube, use the formula $V = s^3$, where s is the measurement of one side.

Example: Find the volume of a cube that measures 5 inches on each side.

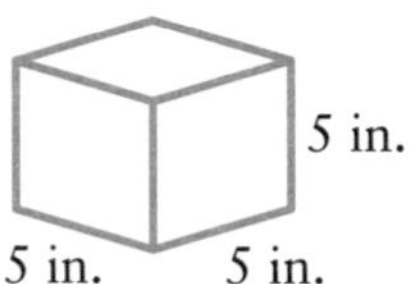

Step 1.	Substitute 5 for s in the formula $V = s^3$.	$V = s^3$ $V = 5^3$
Step 2.	Evaluate the formula.	$V = 125 \text{ in.}^3$

➡ The volume of the cube is 125 cubic inches.

B. Use the formula $V = s^3$ to find the volume of each cube.

5.
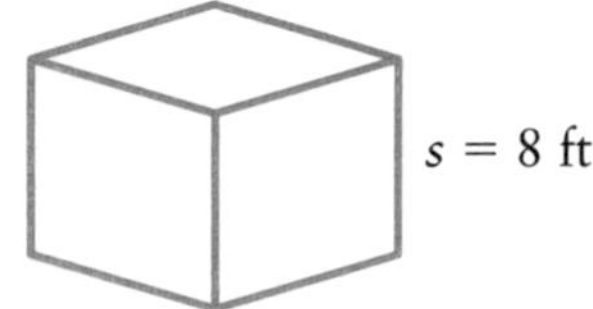
$s = 8$ ft.

6.
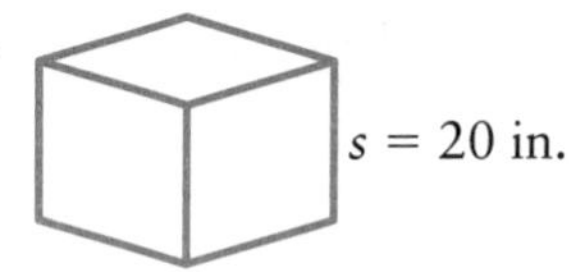
$s = 20$ in.

7.
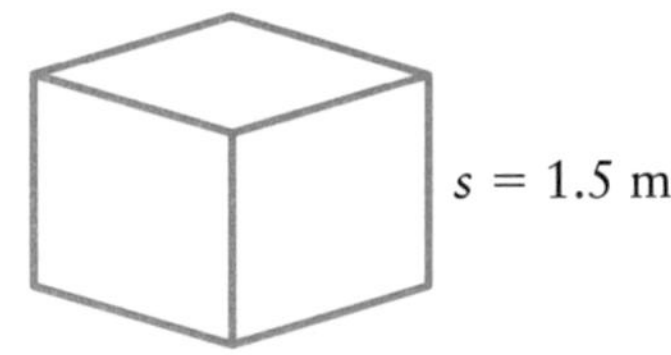
$s = 1.5$ m

8. A cubic yard is a cube that measures 3 feet on each side. How many cubic feet are in one cubic yard?

To find the volume of a cylinder, use the formula $V = \pi r^2 h$, where $\pi = 3.14$, r is the radius, and h is the height.

Example: What is the volume of the cylinder pictured here?

$r = 2$ in.

$h = 5$ in.

Step 1. Substitute 3.14 for π, 2 for r, and 5 for h in the formula $V = \pi r^2 h$.

$V = \pi r^2 h$
$V = 3.14 \cdot 2^2 \cdot 5$
$V = 3.14 \cdot 4 \cdot 5$
$V = 62.8 \text{ in.}^3$

Step 2. Evaluate the formula.

➡ The volume of the cylinder is 62.8 cubic inches.

C. Use the formula $V = \pi r^2 h$ to find the volume of each cylinder.

9.
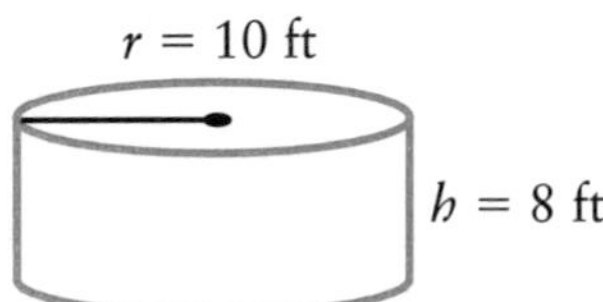
$r = 10$ ft

$h = 8$ ft

10. $r = 3$ in.

$h = 15$ in.

11.
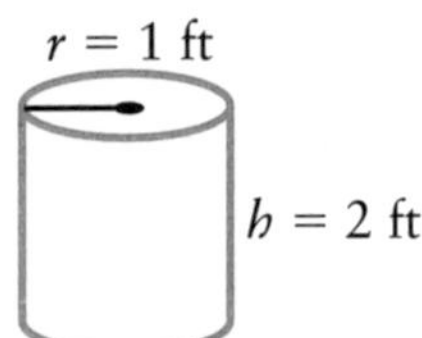
$r = 1$ ft

$h = 2$ ft

To check your answers, turn to page 205.

Similar Figures

Andrea built a small rectangular table for her children. The table was 2 feet wide and 3 feet long. She wants to make a larger table in the same shape for herself. If the larger table is to be 5 feet long, how wide will it be?

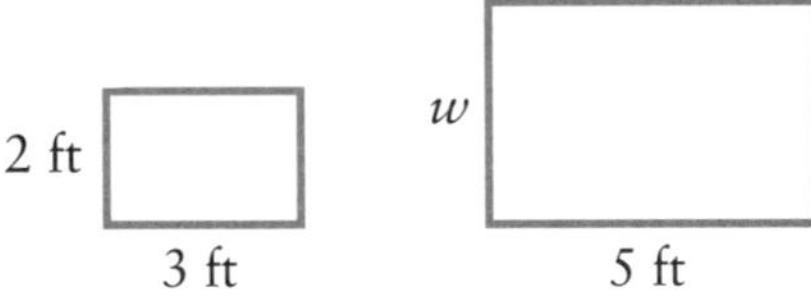

In geometry, two figures are **similar** if they have the same shape. The corresponding sides of similar figures are proportional.

Example: Find the width of the larger table pictured above.

To review solving proportions, turn to page 84.

Step 1. Set up a proportion with ratios of width to length. Let w represent the missing width. $\frac{\text{width}}{\text{length}}\ \frac{2}{3} = \frac{w}{5}$

Step 2. Find the cross products. $3w = 10$

Step 3. Solve for w. $w = \frac{10}{3} = 3\frac{1}{3}$

➡ The new table will be $3\frac{1}{3}$ feet wide.

The corresponding angles of similar figures are equal.

Triangles ABC and DEF are similar because they each have angles of 30°, 60°, and 90°. Notice that *the sum of the angles in a triangle is 180°.*

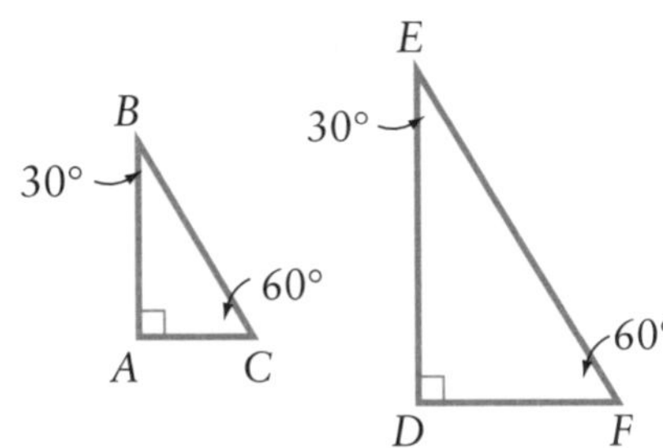

Example: Triangles *RST* and *WXY* are similar. Find the length of side *WY*.

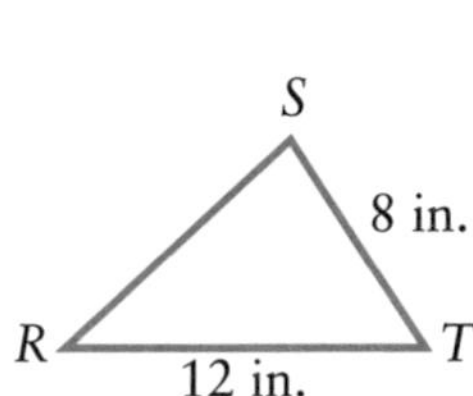

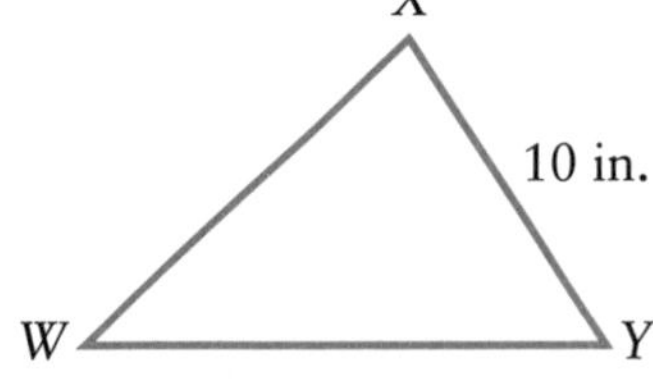

STEP 1. Set up a proportion with ratios of corresponding sides. Let x represent the missing side.

$$\frac{\text{short}}{\text{long}} \quad \frac{8}{12} = \frac{10}{x}$$

STEP 2. Find the cross products.

$$8x = 120$$
$$x = 15$$

STEP 3. Solve for w.

➡ Side *WY* is 15 inches long.

PRACTICE 82

Solve each problem.

1. The two rectangles pictured here are similar. Find the length of the smaller rectangle.

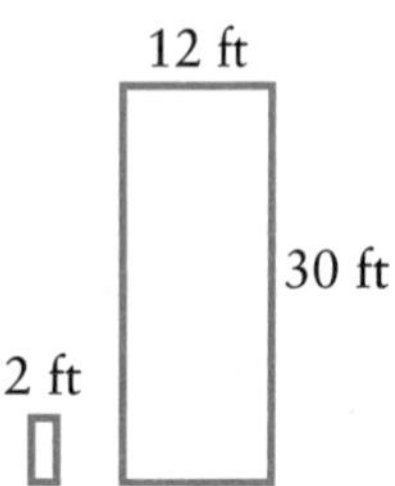

2. The two triangles pictured here are similar. Find the measurement of side *BC*.

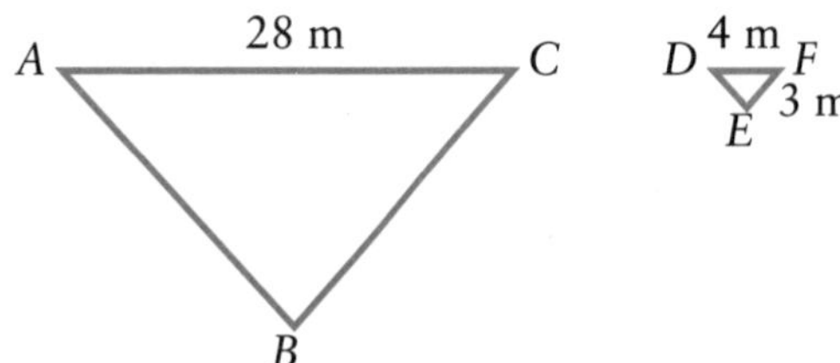

3. A 9-inch-by-12-inch sign was reduced to be 4 inches wide. What is the length of the reduction?

4. The windows in the new Central County office building are different sizes, but their proportions are the same. The ratio of the width to the height of every window is 3:5. The largest window is 9 feet wide. How high is the window?

5. Jennifer wants to enlarge an 8 inch by 10 inch family photo to be poster size. The poster will have the same proportions as the photo. The poster will be 30 inches long. How wide will it be?

6. The diagram shows a pool and a deck. The pool is 20 feet wide and 35 feet long. The deck has the same shape as the pool. If the deck is 48 feet wide, how long is it?

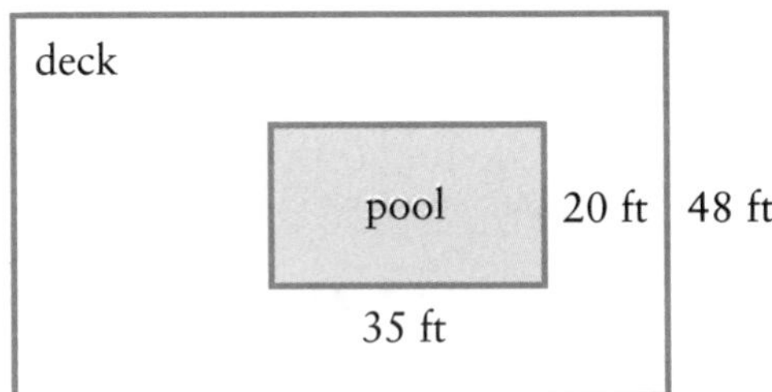

7. Find the value of the missing angles in the triangles below. Are triangles BCD and EFG similar?

To check your answers, turn to page 206.

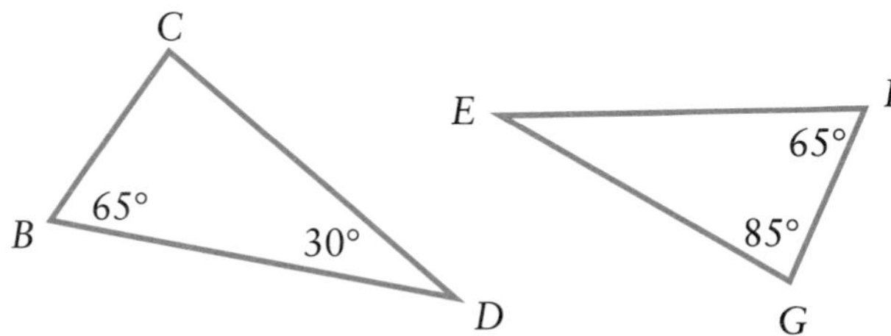

The Pythagorean Relationship

The drawing shows Maxine's backyard. She wants to know the diagonal distance shown by the dotted line.

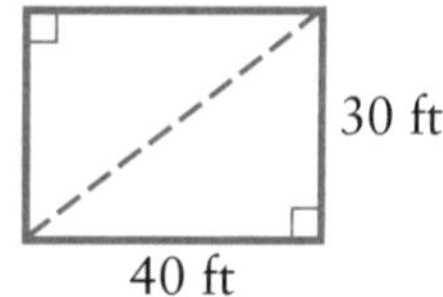

A right triangle is a triangle with a right angle. The dotted line separates Maxine's back yard into two right triangles. The side opposite (across from) a right angle in a right triangle is called the **hypotenuse.** (The diagonal distance across the yard is the hypotenuse of a right triangle.) The other two sides of a right triangle are sometimes called the **legs.**

More than 2,500 years ago, a mathematician named Pythagoras discovered the relationship among the sides of a right triangle. He found that the square of the hypotenuse equals the sum of the squares of the legs.

The formula for the Pythagorean relationship is $\mathbf{c^2 = a^2 + b^2}$, where c is the hypotenuse and a and b are the legs of a right triangle.

Example: Find the diagonal distance for the illustration above.

STEP 1.	Substitute 30 for a and 40 for b in the formula $c^2 = a^2 + b^2$.	$c^2 = a + b^2$ $c^2 = 30^2 + 40^2$
STEP 2.	Evaluate the formula.	$c^2 = 900 + 1{,}600$ $c^2 = 2{,}500$
STEP 3.	To find c, find the square root of 2,500.	$c = \sqrt{2{,}500}$ $c = 50$

➡ The diagonal distance across Maxine's yard is 50 feet.

Example: Find the length of a in the illustration.

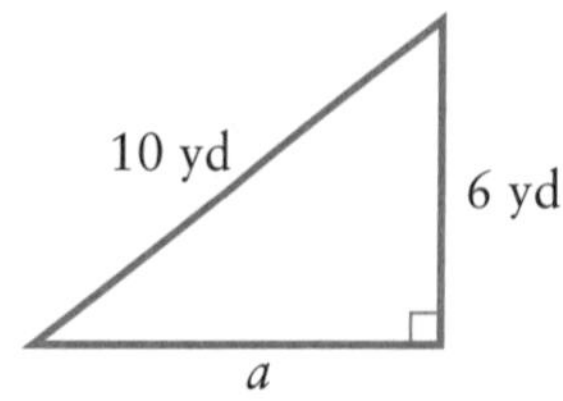

Step 1. Substitute 6 for b and 10 for c in the formula $c^2 = a^2 + b^2$. 10 is the hypotenuse.

$$c^2 = a^2 + b^2$$
$$10^2 = a^2 + 6^2$$
$$100 = a^2 + 36$$

Step 2. Evaluate the formula. Subtract 36 from both sides.

$$64 = a^2$$

Step 3. To find a, find the square root of 64.

$$\sqrt{64} = a$$
$$8 = a$$

➡ The length of side a is 8 yards.

PRACTICE 83

Solve each problem.

1. Find the length of the hypotenuse in the figure below.

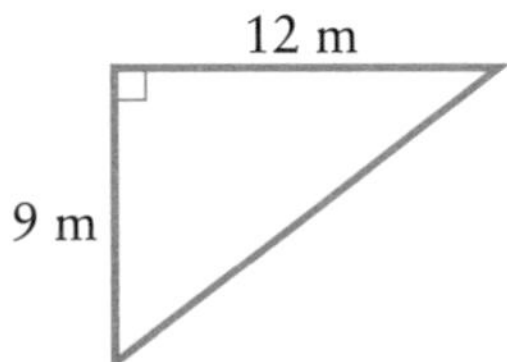

2. The hypotenuse of a right triangle measures 25 meters. One leg measures 20 meters. Find the measurement of the other leg.

3. Find the measurement of diagonal distance BD in the drawing.

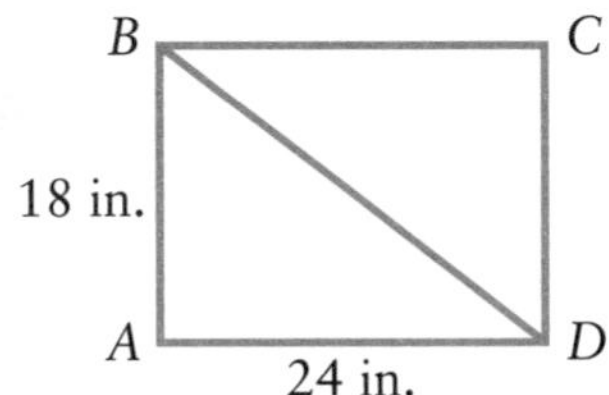

4. Find the length of side b in the illustration.

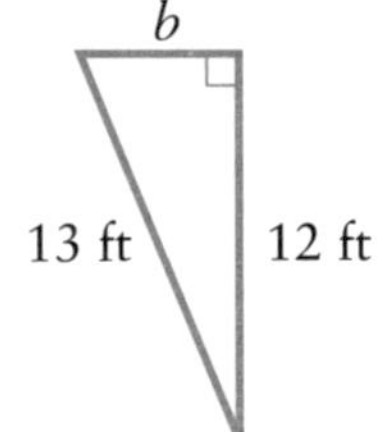

5. What is the distance MO in this drawing?

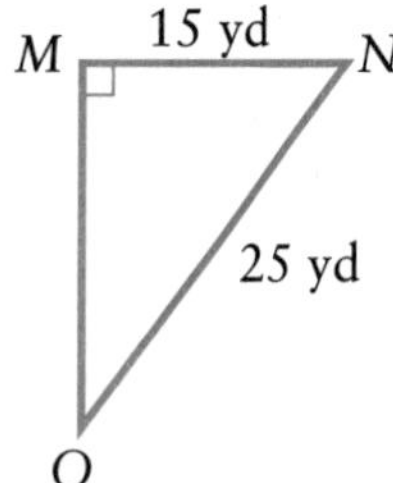

6. The drawing shows Gilda's drive one morning. She drove north 30 miles north from Amesville to Baxter and then 40 miles east to Croydon. What was the shortest distance from her starting point in Amesville to her destination in Croydon?

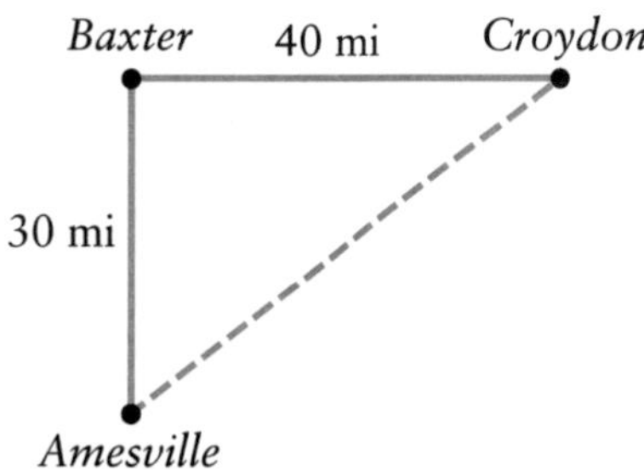

7. A ladder is leaning against a wall. The top of the ladder touches the wall at a spot 12 feet from the ground. The bottom of the ladder is 5 feet from the base of the wall. If the wall is perpendicular to the ground, how long is the ladder?

To check your answers, turn to page 206.

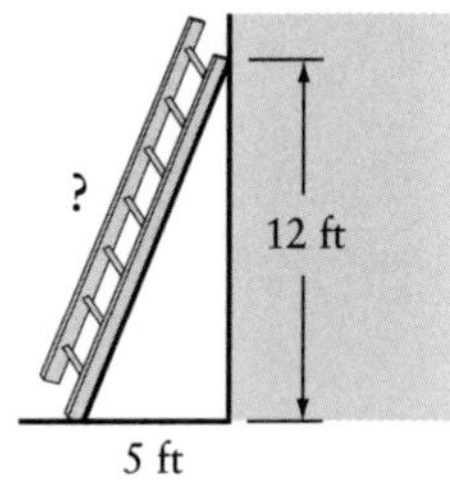

GEOMETRY REVIEW

These problems will help you find out if you need to review the geometry section of this book. Solve each problem. When you finish, look at the chart to see which pages you should review.

For 1–4, identify each angle as acute, right, obtuse, or straight.

1\.

2\.

3\. 60°

4\. 90°

5\. In the illustration, $\angle a = 24°$. Find the measurement of $\angle b$.

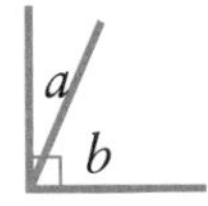

6\. In the illustration, $\angle c = 59°$. Find the measurement of $\angle d$.

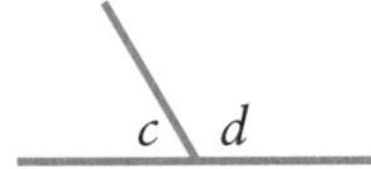

7\. The distance around a circle is called the ______________.

8\. A figure with four right angles and four equal sides is called a

______________.

9\. The three-dimensional figure pictured here is called a ______________.

For problems 10–12, find the perimeter of each figure.

10. 6.5 in. 12 in.

11. $s = 1.8$ m

12.

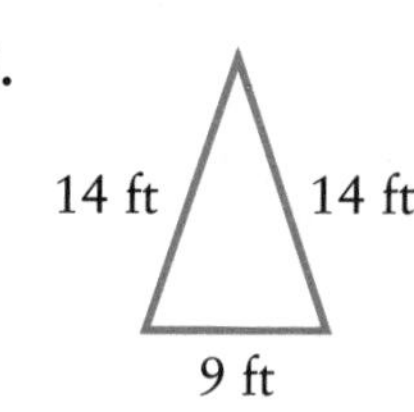

13. Find the circumference of the circle pictured here. Use 3.14 for π.

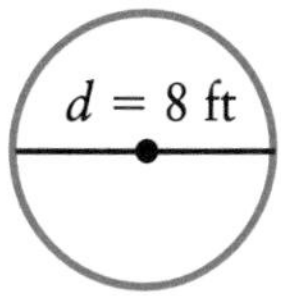

For problems 14–16, find the area of each figure.

14. 7 in. 11 in.

15. $s = 3.6$ m

16. 10 yd 13 yd

17. A square yard is a square measuring three feet on each side. How many square feet are in one square yard?

18. What is the volume of a rectangular storage bin that is 8 feet long, 7 feet wide, and 4 feet high?

19. Find the volume of the cylinder pictured here.

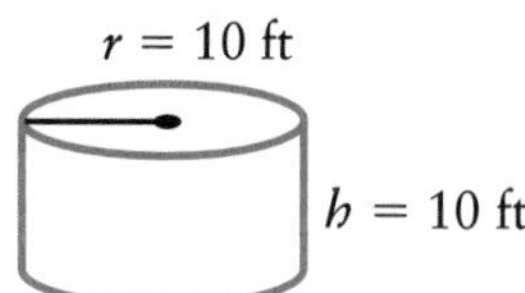

20. The two triangles in the illustration are similar. Find the measurement of side *EF*?

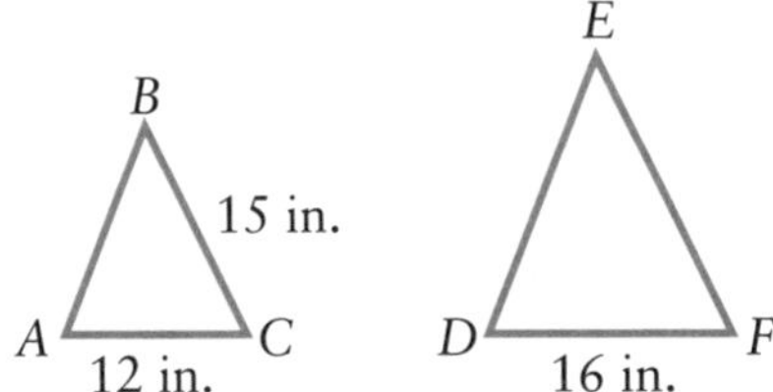

21. What is the diagonal distance *AC* across the rectangular lot pictured here?

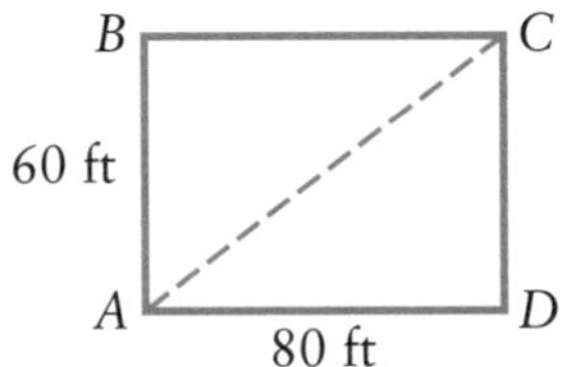

PROGRESS CHECK

Check your answers on page 207. Then return to the review pages for the problems you missed.

If you missed problems	*Review pages*
1 to 4	143 to 144
5 to 6	145
7 to 9	146 to 147
10 to 13	148 to 151
14 to 17	152 to 153
18 to 19	154 to 156
20 to 21	157 to 162

FINAL REVIEW

These problems will help you find out what you need to review. Solve each problem.

1. Write *one million, four hundred seven thousand* in figures.

2. What is 158,641 rounded to the nearest thousand?

3. $265 + 13{,}388 + 5{,}219 =$

4. $46{,}233 - 7{,}585 =$

5. A parents' group is trying to raise $5,000 to send school children on a field trip to the state capital. So far they have raised $3,210. How much more money do they need to raise?

6. $2{,}267 \times 48 =$

7. Fatima makes $17 an hour. How much does she make in a thirty-five hour week?

8. $299 \div 7 =$

9. $5{,}543 \div 82 =$

10. Kirk works as an electrician's assistant. He made $437 on a job that took him 19 hours to complete. What was his hourly wage?

11. Write *twenty-three and fifty-six ten thousandths* in figures.

12. Which decimal is greatest, .32, .309, or .33?

13. Round 2.76 to the nearest tenth.

14. $.124 + 5.62 + 13 =$

15. $6.3 - .184 =$

16. $11 - .566 =$

17. $.0039 \times 54 =$

18. Jordan makes \$12.60 an hour. How much does he make in a week if he works 37.5 hours?

19. $51.8 \div 14 =$

20. $294 \div 4.9 =$

21. Reduce $\frac{15}{35}$ to lowest terms.

22. Change $\frac{51}{8}$ to a mixed number and reduce.

23. Change $8\frac{1}{3}$ to an improper fraction.

24. Which is greater, $\frac{8}{15}$ or $\frac{2}{3}$?

25. $4\frac{1}{3} + 3\frac{3}{20} + 2\frac{5}{6} =$

26. $8\frac{1}{3} - 4\frac{3}{5} =$

27. $\frac{5}{9} \times 12 =$

28. $\frac{3}{5} \times 1\frac{1}{9} =$

29. Jack and Caren want to buy a \$132,000 house. They have to make a down payment of $\frac{1}{5}$ of the price of the house. How much is the down payment?

30. $8 \div \frac{6}{7} =$

31. $6 \div 4\frac{1}{2} =$

32. Jolene paid $9.10 for $3\frac{1}{2}$ pounds of fish. What was the price per pound of the fish?

33. Simplify the ratio 56 to 72.

34. Connie borrowed money from her uncle to pay tuition. So far she has paid back $1,200. She still owes her uncle $1,800. What is the ratio of the amount Connie has paid back to the amount that she borrowed?

35. Write and simplify the ratio of 48 inches to one yard.

36. Solve for n in $\frac{18}{5} = \frac{n}{25}$.

37. Lorenzo drove 130 miles on five gallons of gasoline. If he uses gas at the same rate, how far can he drive on 12 gallons of gasoline?

38. Laila mixed blue paint and white paint in a ratio of 3:2 to paint her house. Altogether she used 15 gallons of paint. How many gallons of blue paint did she use?

39. Change .092 to a percent.

40. Change $\frac{4}{25}$ to a percent.

41. Change 72% to a fraction.

42. 150% of 32 =

43. 10.5% of 200 =

44. Miguel bought a jacket for \$149. He had to pay 5% sales tax. What was the price of the jacket including tax?

45. Find the interest on \$1,500 at 8% annual interest for two years and six months.

46. 72 is what percent of 90?

47. There are 45 bus drivers in Midvale. Of these, 9 drivers are women. What percent of the drivers are women?

48. Sophie got a 12% raise. Her raise is \$42 a week. How much was she making before the raise?

49. $10^2 - 2^3 =$

50. $\sqrt{225} - \sqrt{49} =$

51. $12 - 9 - 11 =$

52. $(-80)\left(-\frac{1}{2}\right) =$

53. $\frac{-20}{24} =$

54. $3 \cdot 18 - 2 \cdot 15 =$

55. Find the value of $2m + 4n$ when $m = 10$ and $n = -3$.

56. Write an algebraic expression for twice the sum of s and eight.

57. Solve for n in $24 = n - 13$.

58. Solve for c in $4c + 7 = 39$.

59. Ten less than half a number is sixteen. Find the number.

60. Solve for y in $6(y - 1) = 48$.

61. In the illustration, $\angle s = 73°$. Find the measurement of $\angle t$.

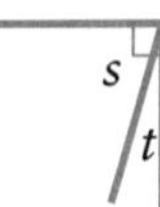

62. The distance from the center of a circle to the circumference is called

the ________________.

For problems 63–64, find the perimeter of each figure.

63.

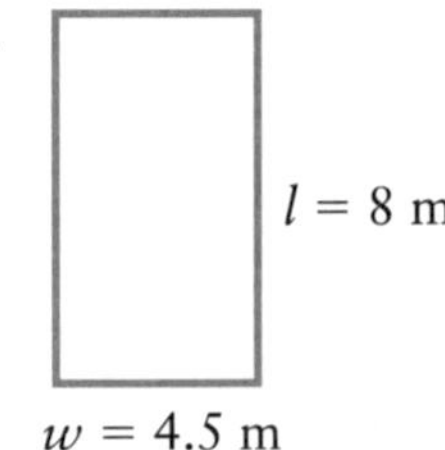

$l = 8$ m

$w = 4.5$ m

64.

$s = 23$ in.

65. Find the circumference of the circle pictured here. Use 3.14 for π.

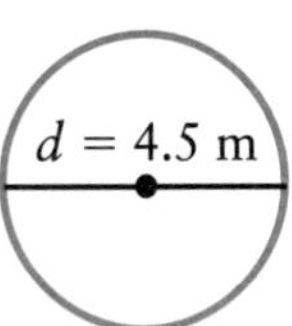

For problems 66–67, find the area of each figure.

66.

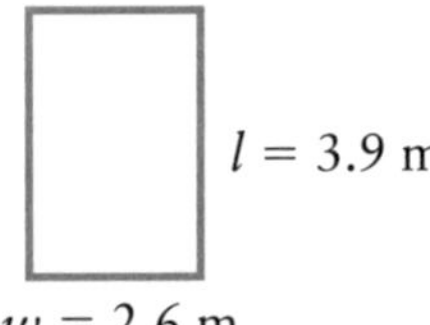

$l = 3.9$ m

$w = 2.6$ m

67.

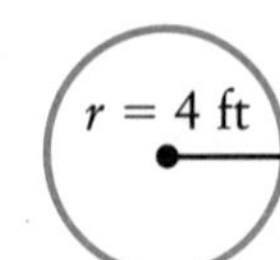

68. Find the length of side XY in this triangle.

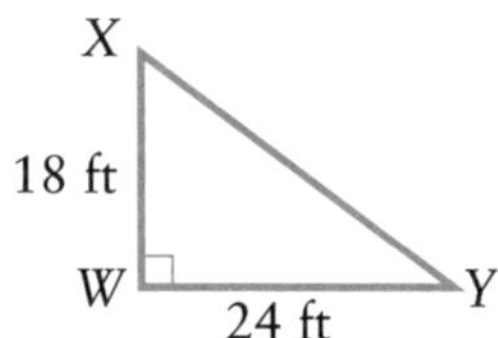

FINAL PROGRESS CHECK

Check your answers on page 208. Circle the problems you missed on the chart below. Review the pages that show how to work the problems you missed. Then try the problems again.

Problem Number	*Review Pages*	*Problem Number*	*Review Pages*	*Problem Number*	*Review Pages*
1	7–9	24	59–60	47	105–107
2	10	25	62–63	48	109–111
3	11–15	26	64–66	49	115–116
4	16–19	27	67–69	50	117
5	16–19	28	70	51	119–121
6	20–21	29	69	52	122
7	20–21	30	71–72	53	124
8	22–25	31	74–75	54	125–126
9	24–27	32	74–75	55	125–126
10	24–27	33	78–79	56	129
11	30–33	34	78–79	57	130–133
12	34	35	80–81	58	134–135
13	34–37	36	84–85	59	134–135
14	38–39	37	84–85	60	138
15	40–41	38	84–85	61	145
16	40–41	39	92–93	62	146
17	42–43	40	94–96	63	148–151
18	42–43	41	94–96	64	148–151
19	44	42	99–101	65	148–151
20	45–48	43	99–101	66	152–153
21	54–55	44	99–102	67	152–153
22	57	45	103–104	68	160–162
23	58	46	105–107		

Answers

Basic Skills Preview
pages 1–6

1. 240,312
2. 30,000
3. 19,269
4. 8,748
5. 35,168
6. 29,241
7. 336 miles
8. 37
9. 240r7
10. 22 months
11. 80.52
12. .54
13. .129
14. 24.316
15. 11.393
16. .2356
17. $10.64
18. 2.9
19. 40
20. $2.90
21. $\frac{1}{9}$
22. $2\frac{2}{5}$
23. $\frac{17}{3}$
24. $\frac{7}{12}$
25. $7\frac{19}{20}$
26. $2\frac{11}{18}$
27. 21
28. $10\frac{1}{2}$
29. $3.90
30. $\frac{1}{24}$
31. $1\frac{1}{4}$
32. $20\frac{1}{4}$ inches
33. 3:4
34. 1:4
35. 2:5
36. $66\frac{2}{3}$
37. 210 miles
38. 114 newcomers
39. 60%
40. 30%
41. $\frac{9}{25}$
42. 108
43. 17
44. $28.90
45. $69
46. 6%
47. 140
48. 25
49. $64 - 25 = 39$
50. $4 + 3 = 7$
51. -21
52. 90 or +90
53. $\frac{3}{4}$
54. $3(5) = 15$
55. 22
56. $n - 15$
57. $w = 12$
58. $4x - 9 = 11 \quad x = 5$
59. $a = 12$
60. $n = 5$
61. 32°
62. 67°
63. 80 in.
64. 42 ft
65. 196 ft^2
66. 162 yd^2
67. 200 ft^3
68. 3 *m*

Place Value
Practice 1
page 8

1.	hundreds	900
2.	tens	80
3.	units or ones	5
4.	thousands	1,000
5.	hundreds	700
6.	tens	60

Reading and Writing Whole Numbers
Practice 2
page 8–10

A.

1. hundred
2. thousand
3. thousand
4. million thousand
5. thousand
6. three thousand, eight hundred
7. nineteen million, seven thousand, two hundred

B.

8. 308
9. 261,000
10. 90,024
11. 4,170,000
12. 804,500
13. 60,300
14. 11,207,000

Rounding Whole Numbers
Practice 3
page 11

1.	40	540	300	5,290	110
2.	900	60,000	4,100	600	4,000
3.	6,000	40,000	3,000	21,000	

Addition with Carrying
Practice 4
pages 12–13

1.	101	102	100	180	144	101	103
2.	103	130	142	112	133	145	130
3.	114	162	141	80	110	73	143
4.	400	313	460	702	485	869	
5.	533	470	1,031	638	350	412	
6.	1,413	1,142	503	1,201	683	1,115	
7.	1,233	1,451	1,570				
8.	2,410	3,039	3,312				
9.	9,020	5,059	6,111				
10.	33,937	70,351					

11. 850 miles
12. 9,049 votes
13. 5,278 pounds
14. $724
15. $30,687

Addition of More Than Two Numbers
Practice 5
pages 14–15

1. 158	171	134	223	150	122	106
2. 147	106	180	205	154	153	153
3. 2,551	4,316	4,290	7,613	9,113		
4. 5,351	3,506	2,554	7,450	2,849		
5. 34,288	96,189					
6. 19,381	20,201					
7. 82,009	31,540					

8. 3,361 tapes
9. 891 calories
10. $152

Subtraction with Regrouping
Practice 6
pages 16–17

1. 29	48	19	39	8	16	28
2. 27	26	9	69	36	17	18
3. 568	154	268	102	186	293	
4. 5,679	6,856	4,649	5,869	2,896		
5. 2,788	2,969	2,786	1,343	1,229		
6. 479	358	587				
7. 2,421	743	3,287				
8. 6,037	29,099	76,559				

9. 177 people
10. 188 miles
11. $5,929
12. 333 people

Regrouping with Zeros
Practice 7
pages 18–19

1. 565	198	115	189	288	328
2. 439	378	94	546	227	38
3. 374	446	311	286	192	24
4. 2,744	662	3,589	50	3,924	3,553
5. 5,730	1,319	4,977			
6. 12,339	10,784	21,118			
7. 71,785	46,552	19,619			

8. $376,740
9. $108,800
10. $1,125
11. 6,670,000 barrels

Multiplication with Carrying
Practice 8
pages 20–21

1. 444	738	296	340	156	198	162
2. 1,368	2,730	1,075	3,024	3,542	3,520	2,574
3. 49,014	31,464	7,756	61,992			
4. 20,784	23,086	35,916	30,552			
5. 1,440	3,750	10,020	7,920			

6. $19,240
7. $112
8. 324 miles
9. 180 inches
10. 192 miles
11. $490
12. $3840

Division by One Digit
Practice 9
page 23

1. 47	23	85	56	44
2. 77	36	45	69	87
3. 79	58	64	88	92
4. 276	839	366	851	928

5. $250
6. $2,760

Division with Remainders
Practice 10
pages 24–25

A.

1. 41r5	39r1	22r6	52r4	78r2
2. 60r3	56r1	44r5	73r4	66r2
3. 93r2	77r5	37r4	99r1	74r1
4. 67r2	81r2	50r3	47r8	65r3

B.

5. 192r6	721r3	304r7	586r2
6. 640r5	817r2	920r6	455r2
7. 272r7	794r1	968r1	582r2

8. 4 pieces
9. 3 feet
10. 4 windows
11. 4 feet

Division by Larger Numbers
Practice 11
pages 27

1. 8	7	6	5
2. 9r30	5r15	4r12	8r15
3. 48	56	65	52
4. 91	77	65	89
5. 73r10	81r25	36r8	26r40
6. 29r60	50r16	82r5	73r15

7. 89 boxes
8. $620
9. $36
10. 35 pounds

Whole Numbers Review
pages 28–29

1. million thousand
2. million thousand

3. 15,206	***4.*** 4,120,008	***5.*** 330
6. 20,000	***7.*** 161	***8.*** 216
9. 7,417	***10.*** 283 miles	***11.*** 2,132
12. 727	***13.*** 41,114	***14.*** 116 members
15. 4,672	***16.*** 20,410	***17.*** 159,354
18. 996 words	***19.*** 93	***20.*** 495
21. 86r15	***22.*** 32 gallons	

Place Value
Practice 12
page 31–32

1. 2	1	0	5	4
2. 8	7	6	3	9
3. 5	9	7	8	2

4. tenths
5. 10
6. hundredths
7. 100

Reading and Writing Decimals
Practice 13
pages 33–34

A.

1. three tenths
2. six hundredths
3. fifteen thousandths
4. four and two tenths
5. eight and seven hundredths

B.

6. .3	*7.* .013	*8.* .02
9. 5.04	*10.* 30.7	*11.* .0012
12. .000016	*13.* 26.9	

Comparing Decimals
Practice 14
page 35

1. .95	.3	.07
2. .4	.04	.061
3. .64	.33	.564
4. .7	.43	.201
5. .302	.505	.77
6. .82	.79	.3303

Rounding Decimals
Practice 15
page 37

1. .4	2.1	.5	8.3	8.0
2. .07	.64	12.50	.42	6.34
3. .105	2.882	.007	.040	4.489
4. 7	3	10	307	42

5. 1.09 yards
6. 1.1 yards

Addition of Decimals
Practice 16
pages 38–39

1. 1.289	1.899
2. 1.7	1.29
3. 2.077	.9917
4. 72.07	10.106
5. 76.746	6.667

6. 56.1°
7. $3.05 million
8. 16 miles
9. 6.3 billion
10. 37,174.1 miles
11. 103.5°
12. 12.75 hours
13. $85.40

Subtraction of Decimals
Practice 17
pages 40–41

1.	5.641	11.65	.218
2.	.282	3.941	.016
3.	.853	5.401	.224
4.	16.11	3.925	5.76
5.	.018	29.2	.251
6.	2.65	2.84	.856

7. .5 million square miles
8. 230.1 million
9. 4,353.4 miles
10. .009
11. .25 meter
12. .6 million
13. .15 meter

Multiplication of Decimals
Practice 18
pages 42–43

A.

1.	5.46	5.784	.6818
2.	33.56	16.34	3.168
3.	18.2	7.38	.24
4.	16.33	17.4	56.32
5.	.42	6.48	131.25

B.

6.	.072	.0188	.00042
7.	.03168	.0254	.0108

8. 81 kilograms
9. $139.50
10. $4.12
11. 9 miles
12. $14.45

Division of Decimals by Whole Numbers
Practice 19
pages 44–45

A.

1.	3.7	.52	3.29
2.	.624	.67	2.8

B.

3. .038	.027	.039
4. .071	.09	.017
5. $.37	$.08	$1.40
6. $.75	$.03	$1.12

7. .63 meter

8. $14.45

Division of Decimals by Decimals
Practice 20
pages 46–47

A.

1. 3.4	.56	7.8
2. 6.3	.58	1.8

B.

3. 2,600	590	3,270
4. 470	60	50

5. 21.4 miles

Division of Whole Numbers by Decimals
Practice 21
pages 48–49

A.

1. 15	40	650
2. 12	250	500

B.

3. 6.9	17.1	16.7
4. 14.29	5.81	1.18

5. 16.7 acres *6.* $2.73

Decimals Review
pages 49–50

1. four thousandths	*2.* eight and one tenth	*3.* .018
4. 1.08	*5.* .52	*6.* 2.4
7. .9172	*8.* 27.96	*9.* 14.5 million
10. 10.491	*11.* 6.248	*12.* 2.274
13. .85 meter	*14.* 4.23	*15.* 6.46
16. 1.634	*17.* $126.65	*18.* 4.8
19. 6.9	*20.* 200	*21.* 2.7 feet

Writing Fractions
Practice 22
pages 52–53

A.

1. $\frac{2}{3}$	$\frac{1}{4}$	$\frac{4}{6}$	$\frac{2}{4}$
2. $\frac{3}{10}$	$\frac{3}{8}$	$\frac{4}{6}$	$\frac{5}{9}$

B.

3. $\frac{3}{4}$	4. $\frac{9}{10}$	5. $\frac{21}{60}$
6. $\frac{127}{1000}$	7. $\frac{19}{36}$	8. $\frac{2}{7}$

Identifying Forms of Fractions
Practice 23
page 54

1. $\frac{6}{7}$	$\frac{4}{5}$	$\frac{8}{200}$
2. $\frac{19}{5}$	$\frac{12}{9}$	$\frac{15}{15}$
3. $8\frac{4}{7}$	$2\frac{3}{20}$	$3\frac{8}{9}$

Reducing
Practice 24
page 55–56

1. $\frac{5}{6}$	$\frac{4}{5}$	$\frac{5}{8}$	$\frac{2}{5}$	$\frac{2}{5}$
2. $\frac{3}{11}$	$\frac{11}{12}$	$\frac{3}{4}$	$\frac{5}{9}$	$\frac{2}{3}$
3. $\frac{4}{7}$	$\frac{3}{5}$	$\frac{5}{9}$	$\frac{1}{2}$	$\frac{1}{12}$
4. $\frac{1}{4}$	$\frac{1}{21}$	$\frac{7}{11}$	$\frac{7}{8}$	$\frac{3}{7}$
5. $\frac{1}{4}$				
6. $\frac{4}{5}$		7. $\frac{1}{7}$		

Raising Fractions to Higher Terms
Practice 25
page 56

1. $\frac{18}{24}$	$\frac{8}{36}$	$\frac{56}{80}$	$\frac{14}{35}$	$\frac{25}{40}$
2. $\frac{54}{63}$	$\frac{3}{36}$	$\frac{30}{50}$	$\frac{24}{54}$	$\frac{16}{22}$
3. $\frac{24}{36}$	$\frac{16}{72}$	$\frac{42}{60}$	$\frac{7}{56}$	$\frac{36}{45}$

Changing Improper Fractions to Whole or Mixed Numbers
Practice 26
page 57

1. $2\frac{1}{2}$	$3\frac{1}{4}$	$5\frac{2}{3}$	$8\frac{3}{5}$	$4\frac{4}{9}$
2. $3\frac{7}{10}$	6	$2\frac{4}{7}$	$5\frac{8}{9}$	8
3. $1\frac{5}{6}$	$3\frac{5}{12}$	5	$2\frac{3}{4}$	$3\frac{2}{7}$

Changing Mixed Numbers to Improper Fractions
Practice 27
page 58

1. $\frac{20}{3}$	$\frac{7}{2}$	$\frac{23}{4}$	$\frac{27}{10}$	$\frac{7}{6}$
2. $\frac{29}{8}$	$\frac{29}{3}$	$\frac{30}{7}$	$\frac{41}{6}$	$\frac{19}{10}$
3. $\frac{25}{3}$	$\frac{29}{12}$	$\frac{37}{4}$	$\frac{39}{8}$	$\frac{27}{7}$

Comparing Fractions
Practice 28
page 60

A.

1. $\frac{3}{5}$	$\frac{5}{12}$	$\frac{8}{15}$	$\frac{7}{9}$
2. $\frac{21}{25}$	$\frac{4}{7}$	$\frac{1}{4}$	$\frac{2}{3}$
3. $\frac{2}{5}$	$\frac{5}{9}$	$\frac{3}{4}$	$\frac{3}{8}$

B.

4. $\frac{3}{8}$	$\frac{5}{12}$	$\frac{3}{4}$
5. $\frac{29}{36}$	$\frac{9}{20}$	$\frac{2}{3}$

Addition of Fractions with the Same Denominators
Practice 29
page 61

1. $\frac{1}{2}$	$7\frac{1}{2}$	$11\frac{1}{2}$
2. $\frac{3}{4}$	$12\frac{5}{6}$	$13\frac{3}{4}$
3. $15\frac{1}{2}$	$7\frac{1}{3}$	$11\frac{3}{10}$

Addition of Fractions with Different Denominators
Practice 30
pages 62–63

1. $1\frac{1}{2}$	$1\frac{23}{72}$	$1\frac{5}{8}$	$1\frac{1}{10}$
2. $1\frac{3}{20}$	$1\frac{7}{24}$	$1\frac{13}{30}$	$1\frac{1}{14}$
3. $1\frac{5}{18}$	$\frac{19}{24}$	$1\frac{7}{30}$	$1\frac{5}{36}$
4. $\frac{23}{24}$	$1\frac{11}{20}$	$1\frac{19}{40}$	
5. $15\frac{11}{24}$	$18\frac{17}{40}$		
6. $9\frac{37}{60}$	$15\frac{7}{18}$		

7. \$$3\frac{3}{4}$ million

8. $12\frac{1}{12}$ feet

9. $12\frac{5}{12}$ hours

10. $21\frac{13}{16}$ pounds

11. $2\frac{1}{4}$ hours

12. $73\frac{1}{4}$ inches

Subtraction of Fractions
Practice 31
page 64

1. $\frac{3}{4}$	$\frac{3}{10}$	$7\frac{5}{12}$	$4\frac{2}{9}$
2. $\frac{3}{5}$	$\frac{11}{21}$	$3\frac{3}{8}$	$2\frac{2}{15}$

3. $53\frac{3}{8}$ inches

4. $7\frac{9}{16}$ inches

Subtraction with Regrouping
Practice 32
pages 66

A.

1. $3\frac{4}{7}$	$1\frac{1}{8}$	$3\frac{2}{9}$	$2\frac{1}{4}$
2. $4\frac{4}{7}$	$3\frac{2}{3}$	$4\frac{3}{5}$	$4\frac{1}{2}$

B.

3. $2\frac{5}{6}$	$4\frac{17}{30}$	$1\frac{5}{12}$	$3\frac{17}{24}$
4. $7\frac{13}{18}$	$1\frac{5}{8}$	$3\frac{7}{18}$	$4\frac{2}{3}$

5. $\$2\frac{1}{8}$ million

Multiplication of Fractions
Practice 33
page 67

1. $\frac{2}{15}$	$\frac{9}{40}$	$\frac{12}{35}$	$\frac{4}{15}$
2. $\frac{5}{48}$	$\frac{21}{40}$	$\frac{8}{45}$	$\frac{1}{56}$

Canceling
Practice 34
page 68

1. $\frac{8}{21}$	$\frac{5}{14}$	$\frac{7}{12}$	$\frac{2}{5}$
2. $\frac{1}{12}$	$\frac{1}{6}$	$\frac{4}{15}$	$\frac{5}{26}$
3. $\frac{2}{27}$	$\frac{4}{5}$	$\frac{12}{35}$	$\frac{3}{28}$
4. $\frac{10}{63}$	$\frac{1}{9}$	$\frac{2}{5}$	$\frac{9}{28}$

Multiplication with Fractions and Whole Numbers
Practice 35
Page 69

1. 6	6	$6\frac{2}{3}$	$1\frac{1}{2}$
2. 8	$8\frac{1}{3}$	4	$1\frac{1}{2}$
3. $1\frac{3}{4}$	$2\frac{1}{2}$	6	$9\frac{3}{5}$

Multiplication with Mixed Numbers
Practice 36
pages 70–71

1. $3\frac{1}{3}$	$1\frac{7}{8}$	$6\frac{1}{4}$	$1\frac{1}{6}$
2. 28	21	$6\frac{1}{2}$	$19\frac{1}{2}$
3. $3\frac{1}{5}$	2	$1\frac{2}{3}$	$2\frac{2}{3}$
4. 4	$1\frac{1}{5}$	5	$2\frac{2}{5}$
5. $1\frac{3}{7}$	$\frac{5}{6}$	$\frac{13}{14}$	$2\frac{1}{10}$

6. 76 ounces **7.** $35.70

Division by Fractions
Practice 37
pages 72

A.

1. $1\frac{1}{5}$	$1\frac{1}{3}$	$1\frac{3}{7}$	$7\frac{1}{2}$
2. $\frac{2}{3}$	$\frac{3}{5}$	$\frac{2}{3}$	$\frac{3}{4}$

B.

3. 16	10	$10\frac{1}{2}$	$7\frac{1}{2}$
4. 8	$3\frac{1}{3}$	8	10
5. $9\frac{1}{3}$	12	$13\frac{1}{3}$	$3\frac{1}{2}$
6. $2\frac{2}{5}$	$5\frac{1}{3}$	5	6

7. 14 containers
8. 18 stakes
9. $4\frac{2}{3}$ or 4 complete aprons

Division of Fractions and Mixed Numbers by Whole Numbers
Practice 38
page 73–74

1. $\frac{1}{10}$	$\frac{2}{45}$	$\frac{1}{16}$	$\frac{1}{12}$
2. $\frac{3}{10}$	$\frac{1}{2}$	$\frac{1}{6}$	$1\frac{1}{5}$

3. $6\frac{1}{4}$ minutes
4. $\frac{7}{10}$ ounce

Division by Mixed Numbers
Practice 39
pages 75

1. $2\frac{1}{2}$	$\frac{2}{3}$	$2\frac{2}{5}$	$4\frac{2}{3}$
2. 4	$3\frac{1}{2}$	$2\frac{1}{4}$	$1\frac{2}{3}$
3. 6	$1\frac{1}{7}$	$1\frac{3}{4}$	$6\frac{1}{4}$
4. $1\frac{1}{2}$	$\frac{5}{6}$	$2\frac{2}{3}$	$1\frac{1}{2}$
5. $\frac{2}{3}$	$4\frac{1}{2}$	$2\frac{1}{2}$	$\frac{4}{5}$

6. $1.80
7. 9 pieces
8. $4.50
9. 6 bookcases
10. 14 rows

Fractions Review
pages 76–77

1. $\frac{13}{1000}$
2. $\frac{4}{11}$ $\frac{3}{8}$
3. $\frac{9}{2}$ $\frac{6}{6}$ $\frac{13}{3}$
4. $4\frac{2}{15}$ $8\frac{4}{7}$
5. $\frac{5}{9}$
6. $\frac{32}{56}$
7. $6\frac{2}{3}$
8. $\frac{29}{6}$
9. $\frac{7}{10}$
10. $\frac{5}{7}$
11. $1\frac{1}{3}$
12. $13\frac{13}{24}$
13. $\frac{13}{30}$
14. $4\frac{7}{12}$
15. $3\frac{1}{2}$
16. $20\frac{3}{8}$ inches
17. $\frac{1}{3}$
18. $5\frac{3}{5}$
19. 6
20. $9,300
21. $\frac{3}{4}$
22. $\frac{7}{24}$
23. $\frac{1}{2}$
24. $3.30

Writing and Simplifying Ratios
Practice 40
pages 79–80

A.

1. $\frac{2}{3}$
2. 5:9
3. 16 to 1

B.

4. 4:5
5. 3 to 4
6. $\frac{4}{5}$
7. 24:2 = 12:1
8. 400:12 = 100:3
9. 1,500:1,000 = 3:2
10. 1,000:1,500 = 2:3
11. $1,500 + $1,000 = $2,500; 1,500:2,500 = 3:5

Ratio of Measurements
Practice 41
pages 81–82

1. 24:60 = 2:5
2. 9:12 = 3:4
3. 4:24 = 1:6
4. 6:14 = 3:7
5. 6:8 = 3:4
6. $\frac{4}{12} = \frac{1}{3}$
7. $\frac{24}{20} = \frac{6}{5}$
8. $\frac{2,640}{5,280} = \frac{1}{2}$
9. $\frac{12}{6} = \frac{2}{1}$
10. $\frac{30}{96} = \frac{5}{16}$
11. $\frac{25}{15} = \frac{5}{3}$
12. 3 to 4
13. 10 to 16 = 5 to 8
14. 2 to 6 = 1 to 3
15. 8 to 4 = 2 to 1
16. 80 to 160 or 20 to 40 = 1 to 2
17. 10:16 = 5:8
18. 2,000:400 = 5:1
19. 800:4,000 = 1:5

Writing Proportions
Practice 42
page 83

1.

$\frac{3}{5} = \frac{9}{15}$
$3 \times 15 = 5 \times 9$
$45 = 45$

$\frac{8}{3} = \frac{16}{6}$
$8 \times 6 = 3 \times 16$
$48 = 48$

$\frac{2}{5} = \frac{10}{25}$
$2 \times 25 = 5 \times 10$
$50 = 50$

2.

$\frac{12}{9} = \frac{16}{12}$
$12 \times 12 = 9 \times 16$
$144 = 144$

$\frac{1}{4} = \frac{20}{80}$
$1 \times 80 = 4 \times 20$
$80 = 80$

$\frac{15}{10} = \frac{3}{2}$
$15 \times 2 = 10 \times 3$
$30 = 30$

3.

$6:24 = 2:8$
$6 \times 8 = 24 \times 2$
$48 = 48$

$20:4 = 30:6$
$20 \times 6 = 4 \times 30$
$120 = 120$

$3:2 = 33:22$
$3 \times 22 = 2 \times 33$
$66 = 66$

4.

$8:28 = 4:14$
$8 \times 14 = 28 \times 4$
$112 = 112$

$3:10 = 9:30$
$3 \times 30 = 10 \times 9$
$90 = 90$

$15:9 = 10:6$
$15 \times 6 = 9 \times 10$
$90 = 90$

Solving Proportions
Practice 43
pages 84–85

A.

1.

$\frac{12}{9} = \frac{8}{n}$
$12 \times n = 72$
$n = 72 \div 12$
$n = 6$

$\frac{2}{7} = \frac{n}{28}$
$7 \times n = 56$
$n = 56 \div 7$
$n = 8$

$\frac{16}{n} = \frac{2}{3}$
$2 \times n = 48$
$n = 48 \div 2$
$n = 24$

2.

$\frac{n}{15} = \frac{4}{6}$
$6 \times n = 60$
$n = 60 \div 6$
$n = 10$

$\frac{3}{5} = \frac{9}{n}$
$3 \times n = 45$
$n = 45 \div 3$
$n = 15$

$\frac{14}{3} = \frac{n}{6}$
$3 \times n = 84$
$n = 84 \div 3$
$n = 28$

3.

$8:11 = 2:c$
$\frac{8}{11} = \frac{2}{c}$
$8 \times c = 22$
$c = 22 \div 8$
$c = 2\frac{6}{4} = 2\frac{3}{4}$

$5:2 = c:7$
$\frac{5}{2} = \frac{c}{7}$
$2 \times c = 35$
$c = 35 \div 2$
$c = 17\frac{1}{2}$

$6:c = 5:3$
$\frac{6}{c} = \frac{5}{3}$
$5 \times c = 18$
$c = 18 \div 5$
$c = 3\frac{3}{5}$

B.

4. $\frac{\text{won}}{\text{lost}}$ $\frac{5}{2} = \frac{15}{n}$

$5 \times n = 30$

$n = 30 \div 5$

$n = 6$ games

5. $\frac{\text{acres}}{\text{bushels}}$ $\frac{6}{180} = \frac{10}{n}$

$6 \times n = 1{,}800$

$n = 1{,}800 \div 6$

$n = 300$ bushels

6. $\frac{\text{wide}}{\text{long}}$ $\frac{3}{5} = \frac{n}{20}$

$5 \times n = 60$

$n = 60 \div 5$

$n = 12$ inches

7. $\frac{\text{sugar}}{\text{butter}}$ $\frac{4}{3} = \frac{3}{n}$

$4 \times n = 9$

$n = 9 \div 4$

$n = 2\frac{1}{4}$ tablespoons

Proportion Shortcuts
Practice 44
pages 86–87

1. $\frac{30}{50} = \frac{m}{35}$

$50 \times m = 30 \times 35$

$m = \frac{\overset{3}{\cancel{30}} \times \overset{7}{\cancel{35}}}{\underset{1}{\cancel{\underset{\cancel{5}}{50}}}}$

$m = 21$

$\frac{9}{400} = \frac{36}{m}$

$9 \times m = 400 \times 36$

$m = \frac{400 \times \overset{4}{\cancel{36}}}{\underset{1}{\cancel{9}}}$

$m = 1{,}600$

$\frac{800}{m} = \frac{400}{15}$

$400 \times m = 800 \times 15$

$m = \frac{\overset{2}{\cancel{800}} \times 15}{\underset{1}{\cancel{400}}}$

$m = 30$

2. $m{:}420 = 4{:}21$

$\frac{m}{420} = \frac{4}{21}$

$21 \times m = 420 \times 4$

$m = \frac{\overset{20}{\cancel{420}} \times 4}{\underset{1}{\cancel{21}}}$

$m = 80$

$8{:}25 = m{:}200$

$\frac{8}{25} = \frac{m}{200}$

$25 \times m = 8 \times 200$

$m = \frac{8 \times \overset{8}{\cancel{200}}}{\underset{1}{\cancel{25}}}$

$m = 64$

$60{:}100 = 48{:}m$

$\frac{60}{100} = \frac{48}{m}$

$60 \times m = 100 \times 48$

$m = \frac{\overset{5}{\cancel{100}} \times \overset{16}{\cancel{48}}}{\underset{1}{\cancel{\underset{\cancel{3}}{60}}}}$

$m = 80$

3. $\frac{\text{miles}}{\text{hours}}$ $\frac{178}{2} = \frac{n}{5}$

$2 \times n = 178 \times 5$

$n = \frac{\overset{89}{\cancel{178}} \times 5}{\underset{1}{\cancel{2}}}$

$n = 445$ miles

4. $\frac{\text{price}}{\text{gallons}}$ $\frac{\$38.40}{3} = \frac{n}{10}$

$3 \times n = \$38.40 \times 10$

$n = \frac{\overset{\$12.80}{\cancel{\$38.40}} \times 10}{\underset{1}{\cancel{3}}}$

$n = \$128$

5. $\frac{\text{inches}}{\text{miles}}$ $\frac{2}{15} = \frac{6}{n}$

$2 \times n = 15 \times 6$

$n = \frac{15 \times \overset{3}{\cancel{6}}}{\underset{1}{\cancel{2}}}$

$n = 45$ miles

6. $\frac{\text{raises}}{\text{total}}$ $\frac{7}{10} = \frac{n}{80}$

$10 \times n = 7 \times 80$

$n = \frac{7 \times \overset{8}{\cancel{80}}}{\underset{1}{\cancel{10}}}$

$n = 56$

7. $\frac{\text{ride by bus}}{\text{total}}$ $\quad \frac{17}{70} = \frac{n}{700}$

$70 \times n = 17 \times 700$

$n = \frac{17 \times \overset{10}{\cancel{700}}}{\underset{1}{\cancel{70}}}$

$n = 170$

8. $\frac{\text{taxes}}{\text{income}}$ $\quad \frac{\$22}{\$100} = \frac{n}{\$10{,}000}$

$100 \times n = 22 \times 10{,}000$

$n = \frac{22 \times \overset{100}{\cancel{10{,}000}}}{\underset{1}{\cancel{100}}}$

$n = \$2{,}200$

9. 1 + 3 = 4 gallons

10. white:total = 1:4

11. $\frac{\text{white}}{\text{total}}$ $\quad \frac{1}{4} = \frac{n}{24}$

$4 \times n = 1 \times 24$

$n = \frac{1 \times \overset{6}{\cancel{24}}}{\underset{1}{\cancel{4}}}$

$n = 6$ gallons

Ratio and Proportion Review
page 88–89

1. 30:36 = 5:6

2. 24 to 16 = 3 to 2

3. $\frac{45}{27} = \frac{5}{3}$

4. 66:22 = 3:1

5. 66 + 22 = 88 games played 66:88 = 3:4

6. 8:12 = 2:3

7. 40:36 = 10:9

8. $\frac{5}{9} = \frac{15}{n}$

$5 \times n = 9 \times 15$

$n = \frac{9 \times \overset{3}{\cancel{15}}}{\underset{1}{\cancel{5}}}$

$n = 27$

9. $\frac{30}{7} = \frac{n}{2}$

$7 \times n = 30 \times 2$

$n = \frac{30 \times 2}{7}$

$n = \frac{60}{7} = 8\frac{4}{7}$

10. $n:8 = 400:50$

$\frac{n}{8} = \frac{400}{50}$

$50 \times n = 8 \times 400$

$n = \frac{8 \times \overset{8}{\cancel{400}}}{\underset{1}{\cancel{50}}}$

$n = 64$

11. $\frac{\text{miles}}{\text{hours}}$ $\quad \frac{115}{2} = \frac{n}{3}$

$2 \times n = 115 \times 3$

$n = \frac{115 \times 3}{2}$

$n = 172\frac{1}{2}$ miles

12. $\frac{\text{defective}}{\text{total}}$ $\quad \frac{3}{500} = \frac{n}{30{,}000}$

$$500 \times n = 3 \times 30{,}000$$

$$n = \frac{3 \times \overset{60}{\cancel{30{,}000}}}{\underset{1}{\cancel{500}}}$$

$$n = 180$$

Writing Percent
Practice 45
pages 91

A.

1. 100% − 75% = 25%
2. 100% − 30% = 70%
3. 100% − 82% = 18%
4. 100% − 12% = 88%

B.

5. 4 × 100% = 400%
6. 3 × 100% = 300%

Percents and Decimals
Practice 46
page 93–94

A.

1. 65%	6%	4.5%	$16\frac{2}{3}\%$
2. 80%	25%	$6\frac{1}{4}\%$	82%
3. 280%	50%	62.5%	400%

B.

4. .55	.08	.125	.02
5. .064	$.33\frac{1}{3}$	.6	2.25
6. .9	.004	.2	5

Percents and Fractions
Practice 47
pages 95–96

A.

1. 70%	60%	$11\frac{1}{9}\%$	18%	
2. 32%	$37\frac{1}{2}\%$ or 37.5%		$33\frac{1}{3}\%$	50%
3. $41\frac{2}{3}\%$	45%	$6\frac{1}{4}\%$	25%	

B.

4.	$\frac{7}{20}$	$\frac{1}{50}$	$\frac{6}{25}$	$\frac{3}{10}$
5.	$\frac{11}{25}$	$\frac{3}{50}$	$1\frac{1}{2}$	$\frac{3}{100}$
6.	$\frac{6}{125}$	$\frac{21}{200}$	$\frac{1}{2500}$	$\frac{11}{400}$
7.	$\frac{1}{8}$	$\frac{5}{6}$	$\frac{3}{7}$	$\frac{1}{12}$
8.	$\frac{9}{10}$	$2\frac{3}{20}$	$\frac{16}{250}$	$\frac{2}{9}$

Common Fractions, Decimals, and Percents
Practice 48
page 97

$50\% = .5 = \frac{1}{2}$

$25\% = .25 = \frac{1}{4}$

$75\% = .75 = \frac{3}{4}$

$12\frac{1}{2}\% = .12\frac{1}{2}$ or $.125 = \frac{1}{8}$

$37\frac{1}{2}\% = .37\frac{1}{2}$ or $.375 = \frac{3}{8}$

$62\frac{1}{2}\% = .62\frac{1}{2}$ or $.625 = \frac{5}{8}$

$87\frac{1}{2}\% = .87\frac{1}{2}$ or $.875 = \frac{7}{8}$

$33\frac{1}{3}\% = .33\frac{1}{3} = \frac{1}{3}$

$66\frac{2}{3}\% = .66\frac{2}{3} = \frac{2}{3}$

$20\% = .2 = \frac{1}{5}$

$40\% = .4 = \frac{2}{5}$

$60\% = .6 = \frac{3}{5}$

$80\% = .8 = \frac{4}{5}$

$10\% = .1 = \frac{1}{10}$

$30\% = .3 = \frac{3}{10}$

$70\% = .7 = \frac{7}{10}$

$90\% = .9 = \frac{9}{10}$

$16\frac{2}{3}\% = .16\frac{2}{3} = \frac{1}{6}$

$83\frac{1}{3}\% = .83\frac{1}{3} = \frac{5}{6}$

Identifying the Percent, the Whole, and the Part
Practice 49
page 98

	%	whole	part
1.	50%	88	44
2.	25%	48	12
3.	90%	$300	$270
4.	75%	20	15

Finding the Part
Practice 50
pages 99–100

A.

1. 36	18	1.8
2. 28	48	84
3. 7	26	9.6

4. $3,900
5. $5.10

B.

6. 48	8	15
7. 11	36	46
8. 63	100	5

9. 27 questions
10. 18 pounds
11. $126

Using Proportion to Find the Part
Practice 51
pages 101–102

1. 15	63	39
2. 27	22.5	72
3. 228	102	1,050

4. $126,000
5. 975 people
6. $9
7. 18 employees

Multi-Step Problems
Practice 52
page 103

1. $476
2. 6 problems
3. $16.38
4. $31.85
5. 15,600 people

Interest
Practice 53
page 104–105

A.

1. $48 $21
2. $48 $21.60

B.

3. $10 $400
4. $16.80 $567

Finding the Percent
Practice 54
pages 105–106

1. 25% 50%
2. 80% $66\frac{2}{3}$%
3. 10% 40%
4. 30% $62\frac{1}{2}$%
5. 30%
6. 10%
7. 60%

Using Proportion to Find the Percent
Practice 55
pages 107

1. 20% $62\frac{1}{2}$%
2. $66\frac{2}{3}$% 35%
3. 10% 50%
4. 90% 125%
5. 9% $62\frac{1}{2}$%
6. 84%
7. 11%
8. 60%

Multi-Step Problems
Practice 56
page 108–109

1. $17.82 − $16.20 = $1.62

   $\frac{\$1.62}{\$16.20} = \frac{1}{10} = 10\%$

2. $63 − $45 = $18

   $\frac{\$18}{\$45} = \frac{2}{5} = 40\%$

3. $1.26 − $1.20 = $.06

   $\frac{\$.06}{\$1.20} = \frac{1}{20} = 5\%$

4. $220 − $187 = $33

   $\frac{\$33}{\$220} = \frac{3}{20} = 15\%$

5. 22 − 16 = 6

   $\frac{6}{16} = \frac{3}{8} = 37\frac{1}{2}\%$

Finding the Whole
Practice 57
pages 110

1.	45	64
2.	28	40
3.	160	150
4.	130	120
5.	42	200

6. 300 members
7. $1,900
8. 60 questions

Using Proportion to Find the Whole
Practice 58
page 111

1.	20	60
2.	175	240
3.	350	400

4. 80 games

Percent Review
pages 112–113

1. 9%
2. .48
3. $41\frac{2}{3}\%$
4. $\frac{17}{20}$
5. 18.75
6. 333

7. 38.4	*8.* 86	*9.* $7,150
10. $31.29	*11.* $80	*12.* $81
13. 60%	*14.* $66\frac{2}{3}\%$	*15.* 25%
16. $62\frac{1}{2}\%$	*17.* 5%	*18.* 15%
19. 190	*20.* 144	*21.* 125
22. 72	*23.* 320,000 people	

Writing Algebra
Practice 59
page 115

1. 8 + 7

2. 9×5 or $9 \cdot 5$ or $9(5)$

3. $\frac{1}{2} + 3$

4. $10 \div 2$ or $2\overline{)10}$ or $\frac{10}{2}$

5. $12 + 3 = 15$

Powers
Practice 60
pages 116–117

A.

1. 100	36	144
2. 64	225	16
3. .09	$\frac{1}{4}$	.0001

B.

4. $36 + 9 = 45$	$64 - 4 = 60$	$1{,}000 - 25 = 975$
5. $64 - 25 = 39$	$16 - 9 + 25 = 32$	$100 - 81 = 19$

Square Roots
Practice 61
page 118

1. $8 + 3 = 12$	$6 - 2 = 4$	$10 - 5 = 5$
2. $4 + 25 = 25$	$81 - 3 = 78$	$12 - 1 = 11$

The Number Line
Practice 62
page 119

1. E	*2.* A	*3.* D	*4.* G
5. B	*6.* C	*7.* F	

Adding Signed Numbers
Practice 63
page 120–121

A.

1. +3	−2	+1
2. 0	−39	+15
3. −25	+49	+3

B.

4. $+9 - 10 + 4 =$
$+13 - 10 = \mathbf{+3}$

$(-3) + (+7) + (-8) + (12) =$
$-11 + 19 = \mathbf{+8}$

5. $-12 + 4 - 3 + 7 =$
$-15 + 11 = \mathbf{-4}$

$(-15) + (+12) + (-2) + (+6) =$
$-17 + 18 = \mathbf{+1}$

Subtracting Signed Numbers
Practice 64
page 122

1. $(-6) - (-10) =$
$(-6) + (+10) = \mathbf{+4}$

$5\ (+12) =$
$5 + (-12) = \mathbf{-7}$

$-13 - (-9)$
$-13 + (+9) = \mathbf{-4}$

2. $(-25) - (30) =$
$(-25) + (-30) = \mathbf{-55}$

$(-21) - (-8) =$
$(-21) + (+8) = \mathbf{-13}$

$32 - (+12) =$
$32 + (-12) = \mathbf{20}$

Multiplying Signed Numbers
Practice 65
pages 123–124

1. −27	−84	32
2. −55	48	90
3. −26	60	23
4. −80	−48	−200
5. 63	0	−108

6. $(+3)(+4) = +12$ or 12 pounds more than now
7. $(+3)(-4) = -12$ or 12 pounds less than now
8. $(-3)(-4) = +12$ or 12 pounds more than now

Dividing Signed Numbers
Practice 66
page 125

1. −5	6	−3
2. $\frac{1}{2}$	10	$-\frac{3}{7}$
3. 9	$-\frac{2}{3}$	1

Evaluating Expressions
Practice 67
pages 126-127

A.

1. $20 - 3 \cdot 5 =$ $20 - 15 = \mathbf{5}$ | $3(5) + 4(9) =$ $15 + 36 = \mathbf{51}$ | $12 - \frac{21}{3} =$ $12 - 7 = \mathbf{5}$

2. $9 \cdot 6 - 5 \cdot 4 =$ $54 - 20 = \mathbf{34}$ | $\frac{50}{2} + 3(10) =$ $25 + 30 = \mathbf{55}$ | $12 + 6^2 =$ $12 + 36 = \mathbf{48}$

B.

3. $2(14 - 3) =$ $2(11) = \mathbf{22}$ | $\frac{9 + 6}{3} =$ $\frac{15}{3} = \mathbf{5}$ | $\frac{12}{1 + 5} =$ $\frac{12}{6} = \mathbf{2}$

4. $(7 - 2)^2 =$ $5^2 = \mathbf{25}$ | $\frac{13 + 8}{7} =$ $\frac{21}{3} = \mathbf{7}$ | $\frac{1}{2}(17 - 3) =$ $\frac{1}{2}(14) = \mathbf{7}$

C.

5. $7 + x = 7 + 3 = 10$ | $c - d = 12 - 5 = 7$

6. $2(x + 4) =$ $2(8 + 4) =$ $2(12) = 24$ | $3a + 4b =$ $3(6) + 4(2) =$ $18 + 8 = 26$

Writing Expressions
Practice 68
pages 128–129

A.

1. $3n$ | $n + 10$

2. $n - 9$ | $\frac{n}{5}$

3. $n - 2$ | n^2

4. $r + 20$ | $z - 7$

5. $8 + w$ | $\frac{t}{15}$

6. $\frac{1}{10}p$ or $\frac{p}{10}$ | **7.** $y - 23$ | **8.** $\frac{w}{35}$

9. $21\% = .21$ $.21g$

B.

10. $\frac{1}{3}(x + 9)$ or $\frac{x + 9}{3}$ | **11.** $2(n - 10)$ | **12.** $\frac{1}{2}(z - 7)$ or $\frac{z - 7}{2}$

13. $\frac{s + 3}{10}$

Using Formulas
Practice 69
pages 130

1. $d = rt$
$d = 60 \cdot 2.5$
$d = 150$ miles

2. $d = rt$
$d = 6.5 \cdot 2$
$d = 13$ miles

3. $d = rt$
$d = 418 \cdot 3.5$
$d = 1{,}463$ miles

4. $d = rt$
$d = 65 \cdot 6$
$d = 390$ miles

5. $d = rt$
$d = 30 \cdot 4.5$
$d = 135$ miles

Understanding Equations
Practice 70
page 131

1. (3)$12 + x = 19$
2. (l)$n - 7 = 15$
3. (4)$30 = \frac{r}{2}$
4. (2)$8y = 50$
5. $s - 13 = 21$
6. $9 = 2z$
7. $p + 6 = 50$
8. $12x = 108$

One-Step Equations
Practice 71
pages 133

1. $\frac{9a}{9} = \frac{72}{9}$
$a = 8$

$41 + 12 = b - 12 + 12$
$53 = b$

$c + 1.5 - 1.5 = 6 - 1.5$
$c = 4.5$

2. $13 \cdot \frac{d}{13} = 2 \cdot 13$
$d = 26$

$\frac{6}{8} = \frac{8e}{8}$
$\frac{3}{4} = e$

$f - 6 + 6 = 19 + 6$
$f = 25$

3. $g + 12 - 12 = 200 - 12$
$g = 188$

$5 \cdot 16 = \frac{h}{5} \cdot 5$
$80 = h$

$\frac{20k}{20} = \frac{300}{20}$
$k = 15$

4. $\frac{9}{18} = \frac{18s}{18}$
$\frac{1}{2} = s$

$t - 2 + 2 = 8 + 2$
$t = 10$

$u + 11 - 11 = 7 - 11$
$u = -4$

5. $n + 7 = 19$
$n + 7 - 7 = 19 - 7$
$n = 12$

6. $16 = \frac{n}{4}$
$4 \cdot 16 = \frac{n}{4} \cdot 4$
$64 = n$

7. $2n = 22$
$\frac{2n}{2} = \frac{22}{2}$
$n = \$11$

8. $n - 28 = 35$
$n - 28 + 28 = 35 + 28$
$n = \$63$

Two-Step Equations
Practice 72
pages 135

1. $8m + 7 - 7 = 55 - 7$

$\frac{8m}{8} = \frac{48}{8}$

$m = 6$

$5a - 3 + 3 = 17 + 3$

$\frac{5a}{5} = \frac{20}{5}$

$a = 4$

$\frac{x}{6} + 3 - 3 = 7 - 3$

$\frac{x}{6} = 4$

$6 \cdot \frac{x}{6} = 4 \cdot 6$

$x = 24$

2. $37 - 9 = 4c + 9 - 9$

$\frac{28}{4} = \frac{4c}{4}$

$7 = c$

$47 + 3 = 10x - 3 + 3$

$\frac{50}{10} = \frac{10x}{10}$

$5 = x$

$1 + 4 = \frac{m}{9} - 4 + 4$

$5 = \frac{m}{9}$

$9 \cdot 5 = \frac{m}{9} \cdot 9$

$45 = m$

3. $3a - 7 + 7 = 23 + 7$

$\frac{3a}{3} = \frac{30}{3}$

$a = 10$

$\frac{1}{2}x + 5 - 5 = 12 - 5$

$\frac{1}{2}x = 7$

$2 \cdot \frac{1}{2}x = 7.2$

$x = 14$

$\frac{n}{15} - 1 + 1 = 3 + 1$

$\frac{n}{15} = 4$

$15 \cdot \frac{n}{15} = 4 \cdot 15$

$n = 60$

4. $3a + 7 - 7 = -5 - 7$

$\frac{3a}{3} = \frac{-12}{3}$

$a = -4$

$13 - 5 = \frac{c}{4} + 5 - 5$

$8 = \frac{c}{4}$

$4 \cdot 8 = \frac{c}{4} \cdot 4$

$32 = c$

$2m - 9 + 9 = 1 + 9$

$\frac{2m}{2} = \frac{10}{2}$

$m = 5$

5. $2m - 5 = 9$

$2m - 5 + 5 = 9 + 5$

$\frac{2m}{2} = \frac{14}{2}$

$m = 7$

6. $3m + 10 = 22$

$3m + 10 - 10 = 22 - 10$

$\frac{3m}{3} = \frac{12}{3}$

$m = 4$

7. $\frac{1}{2}m - 4 = 7$

$\frac{1}{2}m - 4 + 4 = 7 + 4$

$\frac{1}{2}m = 11$

$2 \cdot \frac{1}{2}m = 11 \cdot 2$

$m = 22$

8. $4m + 5 = 65$

$4m + 5 - 5 = 65 - 5$

$\frac{4m}{4} = \frac{60}{4}$

$m = 15$ deliveries

9. $2m - 3 = 37$

$2m - 3 + 3 = 37 + 3$

$\frac{2m}{2} = \frac{40}{2}$

$m = \$20$

Equations with Separated Unknowns
Practice 73
pages 136–137

A.

1. $9m - 3 + 2m - 7 =$
$11m - 10$

$3a - a + 8 =$
$2a + 8$

2. $12 - 7y - 6 + - 5y =$
$6 - 12y$

$13 + c - 6 + 12c$
$7 + 13c$

3. $5a - a = 24$
$\frac{4a}{4} = \frac{24}{4}$
$a = 6$

$15 = 4x + x$
$\frac{15}{5} = \frac{5x}{5}$
$3 = x$

4. $18c + 7c = 50$
$\frac{25c}{25} = \frac{50}{25}$
$c = 2$

$36 = 11c - 2c$
$\frac{36}{9} = \frac{9c}{9}$
$4 = c$

5. $6y - 7 - 5y = 14$
$y - 7 = 14$
$y - 7 + 7 = 14 + 7$
$y = 21$

$9r - 2r + 3 = 31$
$7r + 3 = 31$
$7r + 3 - 3 = 31 - 3$
$\frac{7r}{7} = \frac{28}{7}$
$r = 4$

B.

6. $8e = 30 + 5e$
$8e - 5e = 30 + 5e - 5e$
$\frac{3e}{3} = \frac{30}{3}$
$e = 10$

$3h = 12 + 2h$
$3h - 2h = 12 + 2h - 2h$
$h = 12$

7. $16 - 2m = 6m$
$16 - 2m + 2m = 6m + 2m$
$\frac{16}{8} = \frac{8m}{8}$
$2 = m$

$6z = 10 + z$
$6z - z = 10 + z - z$
$\frac{5z}{5} = \frac{10}{5}$
$z = 2$

8. $5y + 8 = 3y + 26$

$\underline{-3y - 8} = \underline{-3y - 8}$

$\frac{2y}{2} \quad \frac{18}{2}$

$y = 9$

$8n - 9 = 7n + 13$

$\underline{-7n + 9} = \underline{-7n + 9}$

$n \quad 22$

9. $10n = 3n + 7$

$\underline{-3n} = \underline{-3n}$

$\frac{7n}{7} \quad \frac{7}{7}$

$n = -4$

10. $7n + 12 = 2n - 8$

$\underline{-2n - 12} = \underline{-2n - 12}$

$\frac{5n}{5} \quad \frac{-20}{5}$

n

Equations with Parentheses
Practice 74
page 138

1. $3(a + 2)= 18$

$3a + 6 = 18$

$3a = 12$

$a = 4$

$5(n - 3) = 20$

$5n - 15 = 20$

$5n = 35$

$n = 7$

2. $6(y - 5) = 42$

$6y - 30 = 42$

$6y = 72$

$y = 12$

$60 = 3(m + 4)$

$60 = 3m + 12$

$48 = 3m$

$16 = m$

3. $3(r + 4) = 2r + 17$

$3r + 12 = 2r + 17$

$\underline{-2r - 12} = \underline{-2r - 12}$

$r \quad 5$

$9(c - 2) = c + 30$

$9c - 18 = c + 30$

$\underline{-c + 18} = \underline{-c + 18}$

$8c \quad 48$

$c = 6$

4. $6(n - 2) = 18$

$6n - 12 = 18$

$6n = 30$

$n = 5$

5. $7(n + 5) = 56$

$7n + 35 = 56$

$7n = 21$

$n = 3$

Using Formulas Like Equations
Practice 75
pages 139–140

1. $d = rt$

$\frac{160}{2.5} = \frac{r \cdot 2.5}{2.5}$

64 mph $= r$

2. $d = rt$

$\frac{260}{65} = \frac{65t}{65}$

4 hours $= t$

3. $d = rt$

$\frac{40}{16} = \frac{16t}{16}$

2.5 hours $= t$

4. $d = rt$

$\frac{360}{4.5} = \frac{r \cdot 4.5}{4.5}$

80 mph $= r$

5. $c = nr$

$\frac{\$7.20}{2.5} = \frac{2.5r}{2.5}$

$\$2.88 = r$

6. $c = nr$

$\frac{\$38.85}{3} = \frac{3r}{3}$

$\$12.95 = r$

7. $c = nr$

$\frac{\$4.35}{\$2.90} = \frac{n \cdot \$2.90}{\$2.90}$

$1.5 \text{ lb} = n$

8. $c = nr$

$\frac{\$19.20}{12} = \frac{12r}{12}$

$\$1.60 = r$

9. $m = \frac{1}{3}(a + b + c)$

$85 = \frac{1}{3}(78 + 84 + c)$

$85 = 26 + 28 + \frac{1}{3}c$

$85 = 54 + \frac{1}{3}c$

$31 = \frac{1}{3}c$

$93 = c$

10. $m = \frac{1}{3}(a + b + c)$

$44 = \frac{1}{3}(36 + 42 + c)$

$44 = 12 + 14 + \frac{1}{3}c$

$44 = 26 + \frac{1}{3}c$

$18 = \frac{1}{3}c$

$\$54 = c$

Algebra Review
pages 141–142

1. 169
2. $81 - 36 = 45$
3. $7 + 8 = 15$
4. A
5. C
6. B
7. D
8. 7 or +7
9. −2
10. 21 or +21
11. −66
12. 33 or + 33
13. −7
14. $5(5) = 25$
15. $72 - 5 = 67$
16. $3(-2 + 7) = 3(5) = 15$
17. $4(7) - 2(6) = 28 - 12 = 16$
18. $n - 20$
19. $2(s + 12)$
20. $40 + 17 = x - 17 + 17$

    $57 = x$
21. $9 \cdot \frac{m}{9} = 12 \cdot 9$

    $m = 108$
22. $7c - 6 + 6 = 50 + 6$

    $\frac{7c}{7} = \frac{56}{7}$

    $c = 8$
23. $\frac{1}{2}n - 5 + 5 = 9 + 5$

    $2 \cdot \frac{1}{2}n = 14 \cdot 2$

    $n = 28$
24. $9d - 5d = 100$

    $4d = 100$

    $d = 25$
25. $5(m + 2) = 40$

    $5m + 10 = 40$

    $5m = 30$

    $m = 6$
26. $d = rt$

    $192 = r \cdot 4$

    $48 \text{ mph} = r$

Lines and Angles
Practice 76
pages 144–145

A.

1. parallel

2. perpendicular

B.

3. right

4. acute

5. obtuse

6. acute

7. acute

8. obtuse

9. right

10. straight

Pairs of Angles
Practice 77
page 146

1. $90° - 29° = 61°$

2. $180° - 117° = 63°$

3. $125° - 55° = 70°$

4. $90° - 74° = 16°$

Common Geometric Figures
Practice 78
pages 147–149

A.

1. triangle

2. rectangle

3. square

4. square

5. 4

B.

6. circumference

7. $2 \cdot \frac{1}{2}$ ft = 1 ft

8. d. $2r$

C.

9. $\frac{1}{2} \cdot 40$ in. = 20 in.

10. $2 \cdot 4$ ft = 8 ft

Perimeter and Circumference
Practice 79
pages 150–151

A.

1. $P = 2l + 2w$
$P = 2 \cdot 15 + 2 \cdot 7$
$P = 30 + 14$
$P = 44$ m

2. $P = 2l + 2w$
$P = 2 \cdot 42 + 2 \cdot 20$
$P = 84 + 40$
$P = 124$ yd

3. $P = 2l + 2w$
$P = 2 \cdot 7\frac{1}{2} + 2 \cdot 3\frac{1}{2}$
$P = 15 + 7$
$P = 22$ in.

4. $P = 2l + 2w$
$P = 2 \cdot 24 + 2 \cdot 18$
$P = 48 + 36$
$P = 84$ ft

B.

5. $P = 4s$
$P = 4 \cdot 9$
$P = 36$ yd

6. $P = 4s$
$P = 4 \cdot 16$ mi
$P = 64$ mi

7. $P = 4s$
$P = 4 \cdot 6.2$
$P = 24.8$ m

8. $P = 4s$
$72 = 4s$
18 in. $= s$

C.

9. $P = a + b + c$
$P = 15 + 20 + 25$
$P = 60$ m

10. $P = a + b + c$
$P = 16 + 16 + 16$
$P = 48$ ft

11. $P = a + b + c$
$P = 30 + 30 + 41$
$P = 101$ in.

D.

12. $C = \pi d$
$C = 3.14 \cdot 5$
$C = 15.7$ m

13. $C = \pi d$
$C = 3.14 \cdot 12$
$C = 37.68$ in.

14. $C = \pi d$
$C = 3.14 \cdot 3$
$C = 9.42$ ft

Area
Practice 80
pages 152–154

A.

1. $A = lw$
$A = 12 \cdot 9$
$A = 108\ yd^2$

2. $A = lw$
$A = 20 \cdot 16$
$A = 320\ ft^2$

3. $A = lw$
$A = 10 \cdot 7.5$
$A = 75\ m^2$

4. $A = lw$
$A = 16 \cdot 3$
$A = 48\ ft^2$

B.

5. $A = s^2$
   $A = 9^2$
   $A = 81 \text{ ft}^2$

6. $A = s^2$
   $A = 15^2$
   $A = 225 \text{ in.}^2$

7. $A = s^2$
   $A = 1.2^2$
   $A = 1.44 \text{ m}^2$

8. $A = s^2$
   $A = 12^2$
   $A = 144 \text{ in.}^2$

C.

9. $A = \frac{1}{2}bh$
   $A = \frac{1}{2} \cdot 8 \cdot 8$
   $A = 32 \text{ yd}^2$

10. $A = \frac{1}{2}bh$
    $A = \frac{1}{2} \cdot 15 \cdot 12$
    $A = 90 \text{ ft}^2$

11. $A = \frac{1}{2}bh$
    $A = \frac{1}{2} \cdot 7 \cdot 10$
    $A = 35 \text{ m}^2$

D.

12. $A = \pi r^2$
    $A = 3.14 \cdot 10^2$
    $A = 3.14 \cdot 100$
    $A = 314 \text{ in.}^2$

13. $A = \pi r^2$
    $A = 3.14 \cdot 3^2$
    $A = 3.14 \cdot 9$
    $A = 28.26 \text{ ft}^2$

14. $A = \pi r^2$
    $A = 3.14 \cdot 20^2$
    $A = 3.14 \cdot 400$
    $A = 1{,}256 \text{ cm}^2$

Volume
Practice 81
pages 155–156

A.

1. $V = lwh$
   $V = 5 \cdot 4 \cdot 3$
   $V = 60\text{in.}^3$

2. $V = lwh$
   $V = 6 \cdot 5 \cdot 20$
   $V = 600 \text{ ft}^3$

3. $V = lwh$
   $V = 18 \cdot 11 \cdot 3$
   $V = 594 \text{ m}^3$

4. $V = lwh$
   $3{,}840 = 20 \cdot 16 \cdot h$
   $3{,}840 = 320h$
   $12 \text{ in.} = h$

B.

5. $V = s^3$
$V = 8^3$
$V = 512 \text{ ft}^3$

6. $V = s^3$
$V = 20^3$
$V = 8{,}000 \text{ in.}^3$

7. $V = s^3$
$V = 1.5^3$
$V = 3.375 \text{ m}^3$

8. $V = s^3$
$V = 3^3$
$V = 27 \text{ ft}^3$

C.

9. $V = \pi r^2 h$
$V = 3.14 \cdot 10^2 \cdot 8$
$V = 3.14 \cdot 100 \cdot 8$
$V = 2{,}512 \text{ ft}^3$

10. $V = \pi r^2 h$
$V = 3.14 \cdot 3^2 \cdot 15$
$V = 3.14 \cdot 9 \cdot 15$
$V = 423.9 \text{ in}^3$

11. $V = \pi r^2 h$
$V = 3.14 \cdot 1^2 \cdot 2$
$V = 3.14 \cdot 1 \cdot 2$
$V = 6.28 \text{ ft}^3$

Similar Figures
Practice 82
pages 158–159

1. $\frac{\text{length}}{\text{width}}$ $\frac{30}{12} = \frac{x}{2}$
$12x = 60$
$x = 5 \text{ ft}$

2. $\frac{\text{short}}{\text{long}}$ $\frac{3}{4} = \frac{x}{28}$
$4x = 84$
$x = 21 \text{ m}$

3. $\frac{\text{length}}{\text{width}}$ $\frac{9}{12} = \frac{4}{x}$
$9x = 48$
$x = 5\frac{3}{9} = 5\frac{1}{3} \text{ in.}$

4. $\frac{\text{short}}{\text{long}}$ $\frac{3}{5} = \frac{9}{x}$
$3x = 45$
$x = 15 \text{ ft}$

5. $\frac{\text{length}}{\text{width}}$ $\frac{8}{10} = \frac{x}{30}$
$10x = 240$
$x = 24 \text{ in.}$

6. $\frac{\text{width}}{\text{length}}$ $\frac{20}{35} = \frac{48}{x}$
$20x = 1{,}680$
$x = 84 \text{ ft}$

7. $\angle C = 85°, \angle E = 30°$; yes, they are similar.

The Pythagorean Relationship
Practice 83
page 161–162

1. $c^2 = a^2 + b^2$
$c^2 = 9^2 + 12^2$
$c^2 = 81 + 144$
$c^2 = 225$
$c = \sqrt{225}$
$c = 15 \text{ m}$

2. $c^2 = a^2 + b^2$
$25^2 = a^2 + 20^2$
$625 = a^2 + 400$
$225 = a^2$
$\sqrt{225} = a$
$15 \text{ m} = a$

3.
$$c^2 = a^2 + b^2$$
$$c^2 = 18^2 + 24^2$$
$$c^2 = 324 + 576$$
$$c^2 = 900$$
$$c = \sqrt{900}$$
$$c = 30 \text{ in.}$$

4.
$$c^2 = a^2 + b^2$$
$$13^2 = 12^2 + b^2$$
$$169 = 144 + b^2$$
$$25 = b^2$$
$$\sqrt{25} = b$$
$$5 \text{ ft} = b$$

5.
$$c^2 = a^2 + b^2$$
$$25^2 = 15^2 + b^2$$
$$625 = 225 + b^2$$
$$400 = b^2$$
$$\sqrt{400} = b$$
$$20 \text{ yd} = b$$

6.
$$c^2 = a^2 + b^2$$
$$c^2 = 30^2 + 40^2$$
$$c^2 = 900 + 1{,}600$$
$$c^2 = 2{,}500$$
$$c = \sqrt{2{,}500}$$
$$c = 50 \text{ mi}$$

7.
$$c^2 = a^2 + b^2$$
$$c^2 = 5^2 + 12^2$$
$$c^2 = 25 + 144$$
$$c^2 = 169$$
$$c = \sqrt{169}$$
$$c = 13 \text{ ft}$$

Geometry Review pages 163–165

1. acute
2. obtuse
3. acute
4. right
5. $90° - 24° = 66°$
6. $180° - 59° = 121°$
7. circumference
8. square
9. cylinder
10. $P = 2l + 2w$
    $P = 2 \cdot 12 + 2 \cdot 6.5$
    $P = 24 + 13$
    $P = 37$ in.
11. $P = 4s$
    $P = 4 \cdot 1.8$
    $P = 7.2$ m
12. $P = a + b + c$
    $P = 14 + 14 + 9$
    $P = 37$ ft
13. $C = \pi d$
    $C = 3.14 \cdot 8$
    $C = 25.12$ ft
14. $A = lw$
    $A = 11 \cdot 7$
    $A = 77 \text{ in.}^2$
15. $A = s^2$
    $A = 3.6^2$
    $A = 12.96 \text{ m}^2$
16. $A = \frac{1}{2}bh$
    $A = \frac{1}{2} \cdot 13 \cdot 10$
    $A = 65 \text{ yd}^2$
17. $A = s^2$
    $A = 3^2$
    $A = 9 \text{ ft}^2$

18. $V = lwh$
$V = 8 \cdot 7 \cdot 4$
$V = 224 \text{ ft}^3$

19. $V = \pi r^2 h$
$V = 3.14 \cdot 10^2 \cdot 10$
$V = 3{,}140 \text{ ft}^3$

20. $\frac{\text{short}}{\text{long}} \quad \frac{12}{15} = \frac{16}{x}$
$12x = 240$
$x = 20 \text{ in.}$

21. $c^2 = a^2 + b^2$
$c^2 = 60^2 + 80^2$
$c^2 = 3{,}600 + 6{,}400$
$c^2 = 10{,}000$
$c = \sqrt{10{,}000}$
$c = 100 \text{ ft}$

Final Review
pages 166–171

1. 1,407,000
2. 159,000
3. 18,872
4. 38,648
5. $1,790
6. 108,816
7. $595
8. 42r5
9. 67r49
10. $23
11. 23.0056
12. .33
13. 2.8
14. 18.744
15. 6.116
16. 10.434
17. .2106
18. $472.50
19. 3.7
20. 60
21. $\frac{3}{7}$
22. $6\frac{3}{8}$
23. $\frac{25}{3}$
24. $\frac{2}{3}$
25. $10\frac{19}{60}$
26. $3\frac{11}{15}$
27. $6\frac{2}{3}$
28. $\frac{2}{3}$
29. $26,400
30. $9\frac{1}{3}$
31. $1\frac{1}{3}$
32. $2.60
33. 7:9
34. 2:5
35. 4:3
36. 90
37. 312 miles
38. 9 gallons
39. 9.2%
40. 16%
41. $\frac{18}{25}$
42. 48
43. 21
44. $156.45
45. $300
46. 80%
47. 20%
48. $350
49. 92
50. 8
51. −8
52. 40
53. $\frac{-5}{6}$
54. 24
55. 8
56. $2(s + 8)$
57. $n = 37$
58. $c = 8$
59. $\frac{1}{2}x - 10 = 16 \quad x = 52$
60. $y = 9$
61. 17°
62. radius
63. 25 m
64. 92 in.
65. 14.13 m
66. 10.14 m^2
67. 50.24 in.^2
68. 30 ft